CLIMATE CHANGE

Impact on Coastal Habitation

Edited by

Doeke Eisma

Netherlands Institute for Sea Research (NIOZ)
Texel, The Netherlands

LEWIS PUBLISHERS
Boca Raton   Ann Arbor   London   Tokyo

Library of Congress Cataloging-in-Publication Data

Climate change : impact on coastal habitation / editor Doeke Eisma.
                        p.   cm.
              Includes bibliographical references and index.
              ISBN 0-87371-301-X
              1. Environmental sciences. 2. Coastal ecology. 3. Marine ecology. 4. Climatic changes--
Environmental aspects. 5. Sea level--Environmental aspects. I. Eisma, D., Dr.
GE105.C54   1995
333.91′7--dc20   94-35116
                                                                                                                          CIP

© 1995 by CRC Press, Inc.
Lewis Publishers is an imprint of CRC Press

No claim to original U.S. Government works
International Standard Book Number 0-87371-301-X
Library of Congress Card Number 94-35116
Printed in the United States of America 1 2 3 4 5 6 7 8 9 0
Printed on acid-free paper

Introduction

The notion that the world climate is changing because of the release of CO_2 from fossil fuels and other 'greenhouse' gases is widespread. An enhanced greenhouse effect is expected to result in higher mean air temperatures, in changes in the weather, and a higher sea level. The vegetation and life in general on land and in the oceans, as well as many coastal environments, the hydrology of rivers and lakes, erosion and sediment supplied by streams, agriculture, etc. are expected to be affected, with harmful effects for millions of people.

The most comprehensive basis for these expectations are the reports of the Intergovernmental Panel on Climate Change (IPCC[1,2]). According to these reports, mean air temperatures are most likely to increase 2.5°C during the next century with a probable lowest increase of 1.5°C and a highest increase of 4.5°C. Sea level is expected to rise 61 cm, the lowest estimate being 31 cm and the highest 110 cm. These figures apply when no measures are taken to reduce the release of greenhouse gases. There is, however, still considerable uncertainty about what is to be expected. The predictions, based on climate models, because of many interacting factors, can vary widely. In particular, the role of water vapor and changes in cloudiness are not known, and there is no clear relation between the concentration increase of greenhouse gases in the atmosphere and the observed recent increase of the world mean air temperature. It is also uncertain what can be expected to happen at the ice caps of Antarctica and Greenland: more ice melting, or more snow and ice formation.[3,4] World climate has been variable in the past and most probably will continue to be so: increases and decreases in temperature and precipitation will occur because of natural causes. In addition there will also be other man-induced changes besides an enhanced greenhouse effect that may result in changes in the (regional) climate and in (relative) sea level. These add to changes in world climate that are going to happen because of increased concentrations of greenhouse gases in the future: at present no sure estimates of a future climatic change and of changes in living conditions in general, exist.

This is reflected in the contents of this book. In the first chapter C.J.E. Schuurmans, discussing the world's heat budget, makes clear that a distinction between a natural change and man-induced changes is difficult to make and concludes that the present increase in mean air temperature can only in part be attributed to an enhanced greenhouse effect. For the next century a mean air temperature increase of several degrees centigrade can be expected. N.A. Mörner in Chapter 2 stresses the continuing effect of the natural processes that determine sea level worldwide and arrives at a rise of about 20 cm at most during the next century. He also points out, however, that a relatively small rise of sea level may have a strong effect where the coast is low and/ or subsiding.

The other nine chapters of this book are concerned with the effects of climatic change on the coast and on coastal habitation. E.C.F. Bird, in Chapter 3, also discusses sea level and man-induced sea level rise, but the emphasis is on an overview of the effects of a rising sea level on different types of coast. Such effects are difficult to predict, because of delays in the response of the coast to changing conditions of water depth and sediment supply. Regional studies, however, give interesting insights into what can be expected. The flux of water and sediment from rivers to the coast is discussed in Chapter 4 by J.D. Milliman and M. Ren. A global view is given of the water and sediment discharge to the sea, the present-day effects of man's activities and the way river supply may change in response to future anthropogenic changes including climatic change. The flooding of river mouths, in particular by a combination of higher sea level and local subsidence, may result in considerable losses of habitable land, e.g., in Bangladesh and Egypt. Other endangered river mouth areas are in Burma, Vietnam (Mekong Delta), and China.

The response of estuaries to climatic changes is discussed in more detail by K.R. Dyer in Chapter 5. More coastal wetlands have been destroyed by man's activities than by natural processes. A future climatic change will primarily affect estuaries through changes in sediment input and through changes in the transport, erosion, and deposition rates within the estuary. The extent of such changes is very much related to local and regional conditions. The rate of climate change is probably the central factor that determines whether the change is detrimental or not. In Chapter 6 the effects of sea level rise on coastal sedimentation and erosion is discussed by J.D. Wells. There are also problems in predicting these effects because they will differ depending on whether the change is rapid or slow, while short-term effects from natural processes may dominate for a long time. Moreover, there is no universally applicable coastal model: conditions vary from coast to coast. Most rules for coastal response have been developed for sandy shorelines and are not applicable to other coasts; therefore beaches and barrier islands, cliffs, deltas, tidal flats and wetlands, reefs and atolls are discussed separately.

The response of man to the effects of climatic change along the world's coasts is treated in Chapter 7 by F.M.J. Hoozemans and C.H. Hulsbergen. This response consists primarily of coastal defense works against sea level rise and coastal management plans for low-lying coastal areas. These activities are discussed here for different coastal regions in terms of the risks involved as a result of climatic change in relation to the costs of coastal protection. A sea level rise of 100 cm/century (approximately the highest rise predicted in the IPCC reports[1,2]) is taken as a basis, and taking also population growth into consideration, by 2020 about 100 million people will run a serious risk of flooding if no measures are taken for coastal areas. If adequate measures are taken, only 12 million will run such a risk. The largest risks occur along the ocean coasts of south Asia, the southern Mediterranean, the coasts of Africa and the Caribbean and along many small islands, including the Maldives which lie only 1.5 m above present sea level. Along many coasts protection can be provided at relatively low costs in relation to the GNP of the country or countries that are involved, but some coastal areas (mainly along south and southeast Asia and many small islands) remain exposed to high risks when no international help is provided. The sea level rise assumed in this paper is very high: predictions of more likely rates are only half or one-fifth but the results of this assessment are intended as an indication of the relative scale of risks and costs when a serious change in climate occurs during the next century. Low coastal areas, particularly in southern and eastern Asia are exposed to high risks. A special case is coastal cities, which are both causes and victims of climatic change. In Chapter 8, T. Deelstra discusses this with some European cities as examples.

The impact of climatic change on coastal ecology is discussed in Chapters 9 and 10 by V. Noest, E. van der Maarel, and F. van der Meulen, and A. Edwards respectively. The authors in Chapter 9 discuss the impact on the ecology of temperate coastal wetlands, beaches, and dunes; the author in Chapter 10 discusses the impact on the ecology of coastal reefs, mangrove, and tropical seagrass ecosystems. Noest et al. conclude that it is very difficult to predict changes caused by higher air temperatures and higher CO_2 concentrations: possible changes depend very much on the composition of the vegetation and the associated faunas, which are variable and very complex. Also Edwards points out the difficulties involved in making predictions. For coral reefs and sea grass the rate of the sea water temperature rise is probably most important. Mangrove systems are very diverse and may react in very different ways to climatic change, but at present are mostly threatened by cutting and exploitation.

Last but not least, coastal agriculture is affected by climatic change. R. Brinkman points out in Chapter 11 that increased flooding because of sea level rise, an increase in the frequency of storm surges, and an increased salinity of surface water and groundwater, when combined with insufficient coastal protection, are the major hazards. There is also, however, a strong and locally decisive effect of other human activities such as irrigation, pumping of groundwater, river draining, and dam construction.

It follows from the contents of this book that many uncertainties exist, both with regard to the extent of the climatic change itself as with regard to the expected effects on coastal habitation. For coastal (ecological) impact studies the present regional climate scenarios are not considered to be sufficiently reliable.[5] Further uncertainties about the impact on the coast are due to lack of knowledge of the processes involved, of the rates of the changes that are to be expected, and of the response of man. Although there are many uncertainties, the probability of regionally serious changes in climate and sea level, with adverse effects in particular along some low-lying coasts, affecting millions of people, cannot be ignored. This book is intended to give a background of how coastal habitation may be affected.

While putting this book together, I had much help from Dr. S. Jelgersma, The Netherlands, and from a number of anonymous reviewers. This help is gratefully acknowledged.

REFERENCES

1. Houghton, J.T., G.J. Jenkins, and J.J. Ephraums. *Climate Change: The IPCC Scientific Assessment* (Cambridge: Cambridge University Press, 1990), p. 365.
2. Houghton, J.T., B.A. Callander, and S.K. Varney. *Climate Change: 1992: The Supplementary Report to the IPCC Scientific Assessment* (Cambridge: Cambridge University Press, 1992), p. 200.
3. Kellog, W.W. Response to Skeptics of Global Warming, *Bull. Am. Meteor. Soc.* 74(4): 499–511 (1991).
4. Jacobs, S.S. Is the Antarctic ice-sheet growing?. *Nature* 360: 29–33 (1992).
5. Goodess, C.M., and J.P. Palutikhof. Western European regional climate scenarios in a high greenhouse gas world and agricultural impacts. In: J.J. Beukema, W.J. Wolff, and J.J.W.W. Brouns, eds.: *Expected Effects of Climate Change and Marine Coastal Ecosystems* (Dordrecht/Boston/London: Kluwer Academic Publishers, 1990), 23–32.

The Editor

Doeke Eisma, born in Haarlem, The Netherlands, in 1932, majored in physical geography and sedimentology at Utrecht University. In 1961 he went to the Netherlands Institute for Sea Research, Texel, The Netherlands, where he continues to work as Head of the department of Marine Geology and Geochemistry. In 1968 he obtained his Ph.D. at Groningen University with a thesis on the composition and distribution of Dutch coastal sands. He has been studying coastal and shelf deposits, mollusk shells and the transport, deposition and flocculation of suspended matter in different coastal and shelf areas of the world.

In 1970 he became a lecturer in oceanography at Leiden University and from 1980 at Utrecht University, where he became full professor in Marine Sedimentology in 1990. He has published over a hundred research papers, several review papers and chapters, and recently a book entitled *Suspended Matter in the Aquatic Environment*. He also published three popular books (on the Wadden Sea, the North Sea, and the Dutch coast) and numerous articles in newspapers and periodicals on planning, energy, housing, and environmental problems.

Contributors

Eric C.F. Bird
Department of Geography
University of Melbourne
Parksville, Victoria, Australia

R. Brinkman
Soil Resources, Management, and
    Conservation Service
Land and Water Development Division
Food and Agriculture Organization of
    the United Nations
Rome, Italy

Tjeerd Deelstra
The International Institute for the
    Urban Environment
Delft, The Netherlands

Keith R. Dyer
Institute of Marine Studies
University of Plymouth
Plymouth, United Kingdom

Alasdair J. Edwards
Centre for Tropical Coastal
    Management Studies
Department of Marine Sciences and
    Coastal Management
University of Newcastle upon Tyne
Newcastle upon Tyne, United
    Kingdom

Frank M.J. Hoozemans
Harbours, Coasts, and Offshore
    Technology Division
Delft Hydraulics
Emmeloord, The Netherlands

Cornelis H. Hulsbergen
Harbours, Coasts, and Offshore
    Technology Division
Delft Hydraulics
Emmeloord, The Netherlands

Ren Mei-e
Department of Geo and Ocean
    Sciences
Nanjing University
Nanjing, People's Republic of China

John D. Milliman
School of Marine Science
Virginia Institute of Marine
    Science
College of William and Mary
Gloucester Point, Virginia

Nils-Axel Mörner
Department of Paleogeophysics and
    Geodynamics
Stockholm University
Stockholm, Sweden

Vera Noest
Department of Ecological Botany
Uppsala University
Uppsala, Sweden

C.J.E. Schuurmans
Royal Netherlands Meteorological
  Institute
LD Bilt, The Netherlands

Eddy van der Maarel
Department of Ecological Botany
Uppsala University
Uppsala, Sweden

Frank van der Meulen
Department of Physical Geography and
  Soil Science
University of Amsterdam
Amsterdam, The Netherlands

J.T. Wells
Institute of Marine Sciences
University of North Carolina at Chapel
  Hill
Morehead City, North Carolina

Table of Contents

CHAPTER 1

The World Heat Budget: Expected Changes

C. J. E. Schuurmans

INTRODUCTION

Coastal zones, through the ages, have been shaped by the rise and fall of sea level and river flows, which themselves are intimately related to the climate. Climate itself is the result of complex interactions between the atmosphere, oceans, and land, including its ice-covered part. Changes of climate do occur on time scales of years to millennia, part of them being caused by internal feedbacks within the system, external factors remaining constant or nearly constant. Here we may refer to the extremely large change from an interglacial to a glacial climate, which most probably is initiated by relatively small changes in the distribution of solar radiation over the earth.

For other much smaller changes of climate, occurring on time scales of centuries or decades, it is extremely difficult or even impossible to say whether or not external factors are involved. We only know about volcanic dust as a probable cause and, less certain, about intrinsic variations of solar radiation, related to the activity of the sun. In all likelihood, changes of the effective solar radiation, on a global average, are of the order of 1%, which means of the order of some watts per square meter.

Recently, we have become aware that man's activities are no longer irrelevant to climate. Computations show that the changes in atmospheric composition, caused by various antropogenic emissions, may change the radiation balance at the earth's surface by a larger amount than do the above-mentioned variations of the external factors.

Here we first present a global view of the expected changes. The greenhouse effect of the atmosphere is explained and the temperature changes due to increasing greenhouse gas concentrations are estimated, using the global radiation balance

0-87371-301-X/95/$0.00+$.50

concept. It turns out that a global warming of several degrees centigrade and even more is to be expected if man-made emissions of greenhouse gases are not reduced. Uncertainties, however, are present, mainly as a result of the unknown response of world cloudiness to the warming and the equally uncertain role the oceans play in the warming process. Careful analysis of the record of global mean temperature over the past century leads to the conclusion that most probably part of the observed temperature rise must be due to the enhanced greenhouse effect.

Climate changes in most cases are not uniform all over the globe. Even in the case of a uniform forcing, the response of the system may be quite different at different latitudes, and even longitudinally. The enhanced greenhouse effect is a type of forcing that is more or less uniformly distributed over the globe. In order to estimate the latitudinal or 3-D response, climate models based on a heat balance concept (including horizontal transports of heat by atmosphere and ocean) or on a complete set of dynamic and thermodynamic equations governing the atmosphere (the so-called General Circulation Model (GCM)) have to be used. In fact, comprehensive climate studies must use coupled GCMs (atmosphere and ocean).

Here we only describe some of the results of these GCMs. Simulation of the present climate seems to be necessary before any estimation of climate change may be attempted. However, simulation of the present climate is only possible as far as the gross features of climate are concerned. In itself, this is a fantastic result, given the complexity of the system and the relatively short time (25 years) that GCM-climate modeling has been studied. Details of regional climate on scales of hundreds of kilometers or smaller are not reasonably well simulated at present, nor are detailed features of the seasonal variation of climate. We have to bear this in mind when we consider the scenarios of future climate based on GCM computations. These scenarios show large regional differences in the amplitude and rate of change of climate due to the enhanced greenhouse effect. Some of these differences are reasonably well understood and the results may be confidently applied as first guesses in computing possible impacts, e.g., on sea level.

One important feature in this respect, which is not yet treated with confidence by GCMs, is the storm tracks. Any estimation of changes in the characteristics of the storm tracks, therefore, must be looked at very critically.

In general, GCMs show the same range of uncertainty regarding estimated changes of global mean temperature as do the simple radiation balance models. The reasons are exactly the same. More explicit computations of cloud processes and ocean-atmosphere coupling in itself will not narrow the uncertainty gap. Real improvements are only possible on the basis of a better understanding of the processes involved. An example of a process study aimed at improving ocean-atmosphere interaction in GCMs is the incorporation of sea waves in the computation of surface stresses. Inclusion of this process has a strong influence on average pressure distribution (storm tracks).[1]

Finally, in this chapter, some of the direct consequences of the expected climatic changes on sea level are mentioned. The most important aspect of global warming for sea level will be the temperature rise of the ocean. Next to thermal expansion of the ocean, further melting of mountain glaciers may give a significant contribution. The roles of the large ice sheets of Greenland and Antarctica do seem relatively small in the short run, and most probably of opposite sign.

CHANGES OF THE WORLD HEAT BUDGET

The heat budget of our planet as seen from space is very simple. The budget N is made up of only three contributions: one positive (the incoming solar radiation S) and two negative terms (the reflected solar radiation αS and the outgoing longwave radiation I). In formula

$$N = S(1 - \alpha) - I \tag{1}$$

We presume that N on an annual average is nearly zero, so that the earth is in radiative equilibrium. From current satellite measurements we cannot infer this with greater precision than about 5 W/m^2. The solar constant S_0 itself, being about 1360 W/m^2, is known only with an absolute accuracy of ±5 W/m^2. The α value found from measurements of reflected solar radiation is close to 0.3. This means that in a balanced situation the terms in Equation 1 are of the order of 240 W/m^2.

At the earth's surface the global energy balance is a bit more complicated. Here energy terms other than just radiation come in. The budget N_s consists of four terms: the net solar radiation absorbed at the surface S_s, the net flux of longwave radiation I_s, the net fluxes of sensible (H_s), and latent heat (E_s). In formula

$$N_s = S_s - I_s - H_s - E_s \tag{2}$$

Fluxes of sensible and latent heat are mainly upward, but S_s and I_s have large up- and downward components. For S_s it is simply the reflection at the earth's surface which causes a loss of about 10% of the incoming solar radiation. For I_s it is the result of emission of longwave radiation by the earth's surface, being strongly compensated by downward longwave radiation emitted by the atmosphere. The latter component is due to the so-called greenhouse effect of our atmosphere.

In Figure 1–1 the situation is depicted and quantitative estimates of the relevant fluxes are given. Again a state of balance is assumed, which for annual mean conditions will be approximately true. (At any particular time large quantities of heat may be stored in the ocean and/or land surface layer.) Note the strong downward flux of longwave radiation received at the surface. On a global average this flux is even larger than the amount of solar radiation absorbed at the surface!

From the foregoing it will be clear that the earth's atmosphere, through its greenhouse effect, thermally insulates the planet from strong cooling to space. Computations show that a planet without this greenhouse atmosphere, and having the same albedo of 0.3 for solar radiation, would have a radiative equilibrium temperature 33 K below the present temperature at the earth's surface (–18 vs. +15°C). For

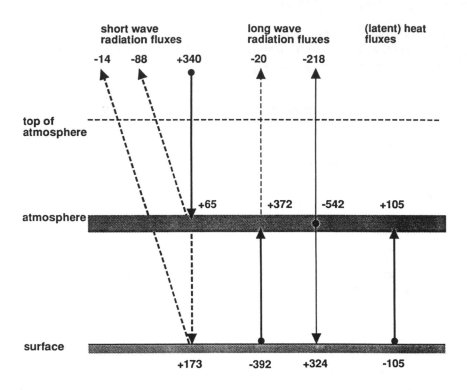

Figure 1–1 The global mean energy balance. Fluxes are given for the earth's surface, the atmosphere, and for the combined system, at the top of the atmosphere. Units: W/m².

the most part the greenhouse effect is due to water vapor (two thirds) and carbon dioxide (nearly one third), with minor additions by other gases like methane, ozone, etc.

As is well known, due to man's activities involving emissions of CO_2 and other greenhouse gases, the concentrations of these substances are increasing, at present at a rate of the order of 1%/year.[2] In view of the above it seems likely that these increases have climatological consequences. More greenhouse gases will cause a better insulation from cooling of the earth to space, which with ongoing absorption of solar energy, will lead to higher temperatures unless the atmosphere or the earth's surface is able to adjust without changing the equilibrium temperature.

A first estimate of the response of global mean surface temperature T_s to a change in the greenhouse effect of the atmosphere can be obtained from simple energy balance models, like the ones given by Equations 1 or 2. Apparently, we have to prefer the model based on Equation 2, since in this case we may expect to get the best estimate of T_s. However, due to uncertainties regarding the physical relation between terms like H_s, E_s, and T_s, we use the "top of the atmosphere" model given by Equation 1.

The idea is that due to increasing concentrations of greenhouse gases, outgoing longwave radiation is progressively more blocked, which will reduce I (planet Earth is getting a positive heat balance). Continued absorption of solar radiation will increase T_s, and, provided I increases with increasing T_s, equilibrium will be restored. In that case the temperature increase required is

$$\Delta T_s = \frac{\Delta I}{\dfrac{\partial I}{\partial T_s}} \tag{3}$$

Clearly, we are in need of a relation $I(T_s)$. Generally, we assume that the outgoing longwave radiation of the earth can be described as a black-body radiation to which the Stefan-Boltzmann law applies

$$I = \sigma T_p^4 \tag{4}$$

Here $\sigma = 5.67 \times 10^{-8}$ W/m² K⁴ and T_p is the planet's radiative equilibrium temperature. The latter may be written as

$$T_p = T_s - \gamma z \tag{5}$$

Here γ is the vertical temperature gradient and z is geometric height. In practice $\gamma \approx 6.5$ K/km while $T_p = 255$ K.

Substituting Equations 5 and 4 in Equation 3 we obtain

$$\Delta T_s = \frac{T_p}{4} \frac{\Delta I}{I} \tag{6}$$

Calculations of radiative transfer show that ΔI amounts to 4.2 W/m² for a doubling of the atmospheric concentration of CO_2 (with a 1% increase per year doubling takes place in about 70 years!).[3]

Using our previous value of I = 240 W/m² we compute $\Delta T_s = 1.12$ K. This is a rather large temperature change compared to the changes of global mean surface temperature observed during the last century. However, it might be unrealistic for two reasons. First of all, equilibrium at the top of the atmosphere does not necessarily mean that there is thermal equilibrium at the earth's surface. Second, in view of a long response time of the earth due to a large heat capacity of the oceans, equilibrium might never be reached.

In the framework of our simple model we will discuss both points, at least qualitatively. For quantitative estimates we have to partly rely upon more complicated models. In order to restore thermal balance at the earth's surface after an increase of surface temperature we may expect at least two things to happen: (1) an increase of evaporation and sensible heat transfer to the atmosphere and (2) some melting of snow and ice. The way these processes affect surface temperature can be positive or negative. In the former case the initial ΔT_s will be enhanced (positive feedback), and in the latter the initial T_s will be reduced (negative feedback). An increase of evaporation and sensible heat transfer will make the atmosphere warmer and its moisture content larger. Water vapor itself is the most important greenhouse gas of the atmosphere, so increasing the amount of water vapor increases the greenhouse back-radiation to the earth's surface, leading to an additional increase of T_s. Clearly, in this case we have a positive feedback. In fact, it is a very strong one, as can be demonstrated on the basis of our simple model. Increasing the water vapor content of the atmosphere means that the effective height from which the longwave radiation escapes into space is increased. In other words, the height z in Equation 5 will increase with T_s. Taking this dependency into account and assuming γ to be constant, instead of Equation 6 we obtain

$$\Delta T_s = \frac{T_p}{4\left(1 - \gamma \frac{\partial z}{\partial T_s}\right)} \cdot \frac{\Delta I}{I}$$

(7)

We easily compute that for a small but realistic value of $\partial z / \partial T_s = 100$ m/K the value of ΔT_s for the situation of $2 \times CO_2$ increases to $\Delta T_s = 3.19$ K.

A second positive feedback, due to the initial increase of surface temperature, is the so-called temperature-albedo feedback. The planetary albedo α consists of two parts: reflection of solar radiation by the atmosphere (mainly by clouds) and by the earth's surface. The first part is very difficult to relate to T_s. In fact, it is one of the remaining uncertainties in estimating the response of T_s to the increasing greenhouse effect. The second part, the relation between T_s and the albedo of the earth's surface, is mainly determined by the temperature-dependent snow and ice cover. Higher T_s reduces the extent of snow and ice, causing a decrease of surface albedo, which, through increased absorption of solar radiation, increases T_s — a positive feedback. Also here some quantification is possible on the basis of our simple model. We start with Equation 1, and retaining the feedback(s) already introduced, we obtain

$$\Delta T_s = \frac{T_p}{4(1 - \alpha)\left(1 - \gamma \frac{\partial z}{\partial T_s}\right) + T_p \frac{\partial \alpha}{\partial T_s}} \cdot \frac{\Delta I}{S}$$

(8)

Using $\alpha = 0.3$, $S = 340$ W/m^2, and a reasonable value of -0.001 K^{-1} for $\partial \alpha / \partial T_s$ we find that with the inclusion of the positive temperature-snow albedo feedback ΔT_s has increased to $+ 4.32$K.

Using more sophisticated models we arrive at essentially the same results: $2 \times CO_2$ means an initial increase of global mean temperature of about 1 K, which, through water vapor and ice albedo feedback, is amplified to about 4 K. Although we have included feedbacks, we still do not know how realistic the equilibrium temperatures are in terms of realization. Using the simple model given by Equation 1 but including a time dependent variation of T_s, given as k dT_s/dt, where k is the heat capacity of the system, we easily derive

$$\frac{d\Delta T_s}{dt} = \frac{(\Delta T_s)_e - \Delta T_s}{\tau} \qquad (9)$$

where e stands for equilibrium and τ is a relaxation time given by τ = Gk, G being defined here as $(\Delta T_s)_e$ = G ΔI. Compare with Equation 8 and note that G includes all feedbacks and numerically amounts to a value of the order 1 m^2 K/W. Note, furthermore that in the time-dependent case ΔI as well as $(\Delta T_s)_e$ are time dependent.

The solution of Equation 9 reads

$$\Delta T_s = (\Delta T_s)_e \, (1 - e^{-t/\tau}) \qquad (10)$$

Assuming that k is determined by an ocean layer 350 m thick we find a value of $\tau = 1 \times 50$ years. This means that the time delay due to the oceans is quite long, but nevertheless 50% of the 4 K warming at $2 \times CO_2$ should be realized in about 30 years after doubling time of CO_2, the remainder of it waiting for realization. Even if the increase of the CO_2 concentration at that time should stop, the earth is committed to the heating until equilibrium is established.

Since the increase of the CO_2 concentration (and of other greenhouse gases) has been going on for more than a century, it is important to analyze existing temperature records in order to see whether or not greenhouse warming is already taking place. In this section we have discussed the global average effect and a comparison with the observed record of global average temperature is therefore appropriate.

In Figure 1–2 this record is shown, updated to 1991. This record is based on surface air temperature data over the continents and measurements of sea surface temperature over the oceans. The data are not uniformly distributed over the globe but several analyses have shown that the temperatures shown are a reliable and representative estimate of the annual averages of global mean surface temperature. Clearly, observed T_s show an upward trend, culminating with the 1980s being the warmest decade and 1990 and 1991 being the warmest years of the whole record. Estimates of observed ΔT_s differ slightly between different analyses (e.g., corrections have to be made for influences like the heat island effect of large cities on some station data), but they all agree to +0.45 ± 0.15°C for the trend over the past century.

How much is this warming in agreement with the expected greenhouse warming? In comparison to the time of starting of the record, around 1860, the earth's atmosphere at present contains an additional amount of greenhouse gases, sufficient to cause an extra greenhouse radiation flux of about 2 W/m^2. (Additional water vapor is not contained in this estimate.) On the basis of our computations and

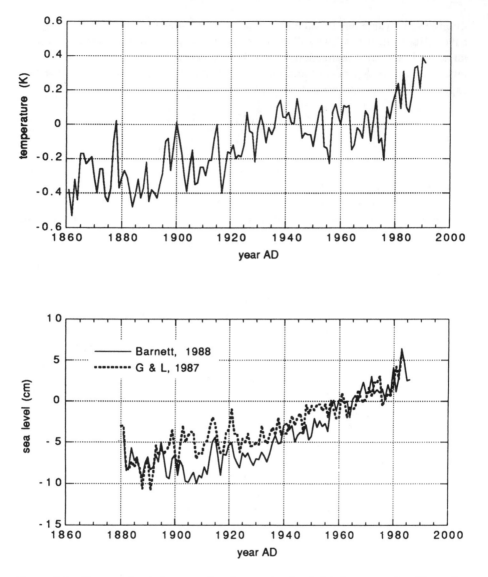

Figure 1–2 Changes of global mean surface temperature[4] and global mean sea level[5,6] over the past century.

taking the ocean delay into account, a global warming of about 1°C might already be expected.

Several reasons have been put forward to explain the apparently lower observed rate of warming over the past century. We mention them here, without a detailed discussion. First, our theoretical estimate of warming might be too large. We mentioned earlier the uncertainties in regard to the feedback of clouds. This and other reasons lead to a large range of uncertainty of $(\Delta T_s)_e$ at $2 \times CO_2$, namely $1.5 < (\Delta T_s)_e < 4.5°C$. Second, factors other than the greenhouse effect might have contributed to the observed record of T_s. The most important of these are changes of solar radiation, volcanic dust, and also man-made aerosols (particles) in the atmosphere. Another important factor is the internal interaction within the climate system, e.g., between the atmosphere and the oceans (see the next section).

Because of this large number of intervening factors, it is impossible to conclude what part of the observed heating is caused by the greenhouse effect, at least not from a comparison of global mean estimates alone. This attribution problem can only be overcome when observed warming rates are larger than might be expected from any of the factors, except the greenhouse effect.

As shown in Figure 1–2, trends in other climate parameters related to temperature are also observed. Globally averaged sea level has risen by 10 cm from 1880 to the present. This rise, to a large extent, is due to expansion of the seas and oceans, which show a temperature increase in association with the atmospheric warming over the last century. An equally large part of the observed sea level rise is due to melting of glaciers and small ice caps. The contribution of the large ice caps of Greenland and Antarctica is probably small.

WORLDWIDE PATTERNS OF CHANGE

Even for the layman it will be evident that global changes of climate are not necessarily globally uniform. Experience shows that temperature changes and temperature variability in general increase as a function of latitude. For instance, the global warming in the first half of this century, as shown in Figure 2–2 varied from 0.2°C in the tropics to 2.2°C in the polar cap (north of 65° N). The reasons for the latitudinal amplification are reasonably well understood (see later discussion) and we may expect this phenomenon to occur even in case of a uniformly distributed cause of climate change. Apart from latitudinal differentials, large differences may occur between places on the same parallel. These differences are mainly the result of surface conditions, orography, and location with respect to the oceans. It will be clear that simple concepts, such as the radiation balance at the top of the atmosphere, are no longer sufficient to study the behavior of the climate system at any particular place. As a matter of fact, at any one place over the globe, radiation balance does not exist. At low latitudes we usually find that more solar radiation is going into the system than is emitted by it in the form of infrared radiation. At higher latitudes we find the opposite. Clearly, in order to maintain a thermal equilibrium at these places heat has to be transported from lower to higher latitudes, and this is exactly what makes regionally differentiated climate models so complicated.

In order to solve the problem, climate researchers have adopted the approach used by meteorologists to make their weather forecasts. Changes of weather in most instances are caused by the appearance, displacement, and development of weather systems, external conditions like incoming solar radiation remaining almost equal.

These weather systems, like cyclones and anticyclones, are exactly the flow phenomena causing the transports of heat necessary to restore and maintain the thermal
balance, at least in the long-term climatological sense. Atmospheric models used for
weather forecasting are therefore ideally suited to make regionally differentiated
computations of climate.

In order to appreciate the results that are given later, imagine the atmospheric
model to consist of columns of air, bounded by the earth's surface (or ocean surface)
and some upper level at a height of some 40 km, where the air is already extremely
thin. The columns have sides of some 100×100 km and cover the whole globe. In
the vertical direction the columns are subdivided into boxes, some tens in number,
with the height being smallest near the surface where the air is most dense.

For such an atmosphere consisting of boxes of air it is possible to compute for
each box what will happen with temperature, pressure, wind, and atmospheric
composition (e.g., moisture content) using a small number of physical laws (equations of motion, first law of thermodynamics, etc.). In weather forecasting these
computations have to start from some observed state of the atmosphere, but for
climate applications the initial conditions may be purely hypothetical (e.g., a state of
complete rest with a temperature that is the same everywhere). It is hard to believe,
but it has been empirically proven many times, that out of such an artificial state an
atmosphere of the type we know evolves (with warm tropics and cold poles, with
cyclones and anticyclones at midlatitude, etc.), given, of course, the right boundary
conditions. These include a geography with continents, oceans, and mountains, a
specification of surface roughness, thermal conductivity, and moisture content of the
surface, etc. Most importantly, the amount and distribution of solar radiation at the
top of the atmosphere has to be given, and, if not computed by the model, also the
composition of the atmosphere (e.g., the concentration of CO_2).

Simulations of the present climate by such models (GCMs) are still not perfect,
but the gross features of the earth's climate are reasonably well simulated. It should
be remarked, however, that this does not mean that we understand the present climate
in all aspects. GCMs are extremely complicated models (comparable in complexity
to the atmosphere itself) and the successful simulation of a certain feature, e.g., the
cyclone track in the North Atlantic, does not mean that we understand why it is
located at about $50°$ N in the West Atlantic and some $5°$ more northerly in the East
Atlantic. We have to accept it as a result of a complicated chain of events and use
the GCM as an experimental tool. GCM experiments, although of recent date
(earliest attempts around 1965, but with increasing computer power long climate runs
only possible from about 1985), have revolutionized our view of climate change and
variability. More and more it has become clear that the atmosphere itself, in its
interaction with the underlying earth's surface, may give rise to long-term changes
without a substantial change of external forcing conditions like incoming solar
radiation or the concentration of greenhouse gases like CO_2. It has been shown, for
instance, that with external forcing kept essentially constant, global mean surface
temperature may change by nearly the same amount as observed over the last
century. It must be stressed, however, that in this case the atmospheric GCM was

coupled to an ocean GCM, which extends the range of internal interactions of the climate system, especially on the larger time scales.

From such experiments, a valid conclusion seems to be that most, if not all, variability of climate observed in the real atmosphere is the result of varying interactions between parts of the system. If this is the case for a parameter such as global mean temperature, it must be even more true for any parameter of local scale.

What does this mean for the possibility of changing the state of climate by external factors? From simple balance considerations we, for instance, concluded that doubling the CO_2 content of the atmosphere will result in a global warming of several degrees centigrade. What happens in a GCM experiment with $2 \times CO_2$? Fortunately, we arrive at the same result: running a GCM with $2 \times CO_2$ until equilibrium is reached and comparing its global mean temperature with that of a control run with $1 \times CO_2$ shows a global warming of 1.5 to 5°C, depending on the GCM used.[7] Apparently, equilibrium climates of $2 \times CO_2$ and $1 \times CO_2$ differ more from each other than the climates resulting from varying internal interactions, at least on a global scale. On a global average we may thus expect that with ongoing increases of atmospheric CO_2 content, we sooner or later will reach a temperature level well above the level of climates at $1 \times CO_2$.

In this section however, we are also interested in the worldwide pattern of change. What do GCMs tell us about the regional changes of climate at $2 \times CO_2$? As far as surface temperature is concerned the equilibrium climate at $2 \times CO_2$ is everywhere warmer than the climate $1 \times CO_2$, with the tropics showing the least and the polar areas the largest warming. The larger warming in the polar areas has been analyzed carefully and at least four processes are identified which are responsible for it:

1. In a warmer climate sea ice is formed later in autumn and remains thinner throughout winter, causing the heat flux from the ocean to the atmosphere to be larger.
2. Less ice and snow decreases the surface reflectivity (albedo), though which a larger amount of solar radiation can be absorbed.
3. A warmer global atmosphere contains more moisture, so more moisture will condensate in the polar areas, releasing more latent heat.
4. In the polar areas any additional heat in the lower layers of the atmosphere is trapped in the lower layers, because of the stable stratification, inhibiting convection into the upper atmosphere.

Equilibrium climate at $2 \times CO_2$ is something artificial, which might be established neither globally (because of ongoing increase of CO_2 concentration), nor locally (because of different delay times, depending on ocean influence). For this reason, a more sensible thing to look at is the temperature change realized at the time of $2 \times CO_2$, as compared to $1 \times CO_2$.

The pattern shown in Figure 1–3 has been computed by running a coupled atmosphere/ocean GCM twice, one time starting from initial conditions comparable to the present and increasing the CO_2 content of the atmosphere by 1%/year until the concentration is doubled, the second time starting from the same initial conditions and running the model for the same length of time (about 70 years) without changing

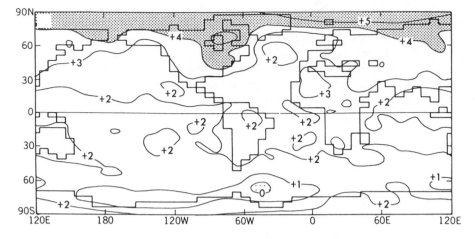

Figure 1–3 Estimated difference (in °C) of surface air temperature between a climate of $2 \times CO_2$ and $1 \times CO_2$: geographical distribution of annual average response.[8]

the CO_2 content (or any other external factor). The averages of the last 20 years or so, of both runs are subtracted from each other and the difference is taken as the effect or climate change caused by the increasing CO_2 content of the atmosphere. Clearly, with a realistic 1%/year increase of the greenhouse effect the earth's climate at the time of $2 \times CO_2$ (some 70 years from now) will be warmer than the present climate nearly everywhere. Warming is greater over the continents than over the oceans, while the enhanced warming at the polar latitudes, as expected from earlier equilibrium experiments, is still present. Relatively low values of warming are found in the southern oceans and in the northern North Atlantic. In these areas strong vertical mixing takes place, spreading the available heat over a much thicker layer of water than elsewhere, thereby reducing the CO_2-induced warming at the sea surface. Other near-surface processes, such as enhanced fresh water flux from the increased precipitation, reducing the northward ocean transport of heat, contribute to the lower sea surface temperature. At any rate, we may conclude that the greenhouse warming will be a worldwide phenomenon, which at the time of $2 \times CO_2$ reaches a magnitude larger than the changes caused by natural climate variability, certainly over the continents and in the polar areas.

In Figure 1–3 it is the change of annual mean temperature that has been displayed. Since in the coupled GCM runs the seasonal cycle was included, climate differences between $2 \times CO_2$ and $1 \times CO_2$ can also be studied for the seasons separately. As shown in Figure 1–4, at most latitudes the warming does not show a strong seasonal variation, except for the polar areas. Note, however, that in this figure only the latitudinal averages of the temperature change are given. Apparently, the enhanced warming at the north polar latitudes is only a phenomenon of the winter half of the year. This is in agreement with the explanation given earlier.

It is tempting, of course, to compare the present worldwide warming pattern with the pattern shown in Figure 1–3. The agreement between the two, however, is small, which means that we cannot yet conclude that the present warming is due to the

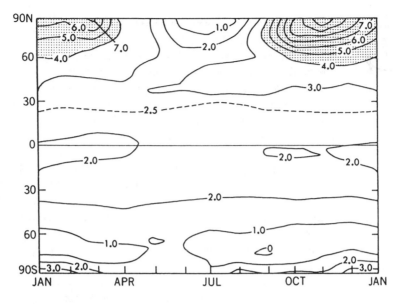

Figure 1–4 Estimated difference (in °C) of surface air temperature between a climate of $2 \times CO_2$ and $1 \times CO_2$: seasonal variation of zonal average response.

enhanced greenhouse effect. It should be realized, however, that the observed pattern includes the changes due to internal climate variability, which has been until now, and perhaps will be for many years to come, of the same or larger magnitude than the presently realized greenhouse warming signal. Although we cannot yet use the geographical pattern as a fingerprint of greenhouse warming, the seasonal distribution of global average warming may give some indication. For the recent decade (1981 to 1990) we find that in all seasons the climate was warmer than the 1951 to 1980 reference climate. The anomalies of global temperature are the following:

Winter (December to February)	+0.26°C
Spring (March to May)	+0.24°C
Summer (June to August)	+0.19°C
Autumn (September to November)	+0.17°C

Such a uniform distribution of warming, with a tendency to be larger during the northern winter, is what is to be expected on the basis of Figure 1–4. However, climate changes due to factors other than the varying greenhouse effect may show this signature as well. In other words, it is still not possible to attribute the present warming to the increased greenhouse effect.

THREATS TO HABITATION IN THE COASTAL ZONE

Storms and floods are among the heaviest threats to coastal zone habitation. Prospects of increasing numbers or intensities of storms and rise of sea level are therefore of greater importance than other aspects of climate change to coastal areas. What can we say about this in relation to the changes discussed in the foregoing sections?

As far as storms are concerned, again we have to rely on the GCM experiments. First, it must be remarked that until now simulations of storm tracks of extratropical cyclones for the present climate have not been completely successful. In the simulations, cyclone activity in general is somewhat lower than in the real atmosphere, while Atlantic depressions on average move too far into Europe before they disappear. The reasons for these systematic errors are being studied.

In most $2 \times CO_2$ climate runs, as compared to the single CO_2 runs, both in the Pacific and the Atlantic a slight polar shift of the main storm tracks is observed. Changes in intensity are as yet inconclusive. The often-aired statement that the reduced meridional temperature gradient will be associated with decreased storm intensities is probably too simple. In wintertime land-sea temperature contrasts may indeed become smaller than they are at present, which will reduce cyclone intensities. On the other hand, an increased moisture content of the warmer air gives rise to a larger release of latent heat in depressions, which enhances storm activity.

Our knowledge regarding the behavior of tropical cyclones in a warmer climate is even more scanty. Observational experience that tropical storms (hurricanes, typhoons) only form where the sea surface temperature exceeds 27°C might lead to the expectation that in a warmer climate the frequency of occurrence of such disturbances will increase. However, factors other than sea surface temperature are also of importance and it is not certain that theoretical studies are able to take them properly into account. GCMs, on the other hand, are also not properly suited to study the problem. Tropical cyclones happen to be of horizontal dimensions that are smaller than the size of air columns forming the basic computational elements of GCMs. This means that the occurrence or nonoccurrence of tropical cyclones has to be described in terms of larger scale parameters than are resolved by the model. Despite these difficulties, it is found that the simulated distribution of tropical storms compares well with the observed patterns. Results for a $2 \times CO_2$ climate are as yet somewhat contradictory. In one GCM it was found that the number of tropical disturbances increased by 50%, with little change in average intensity, while in another $2 \times CO_2$-model run the number of storm days decreased by 10 to 15%.[3]

Theoretical arguments strongly suggest that with higher sea surface temperatures the maximum possible intensity of tropical cyclones will increase (40% increase of wind power, i.e., square of wind speed, for an increase of 3°C in sea surface temperature). However, in the present climate the maximum possible intensity is very seldom reached. Furthermore, as far as observations are concerned, no systematic trend in the number of tropical cyclones is found, neither over the last century, nor over the more recent decades with better monitoring facilities.

The relation between global mean temperature and the height of sea level is much clearer than its association with world storminess discussed above. Global warming as being the likely result of the enhanced greenhouse effect will cause a worldwide rise of sea level due to the expansion of seawater at higher temperatures and the melting of glaciers. On a global scale this rise of sea level may be counteracted by an increased accumulation of ice on the large ice caps of Greenland and Antarctica, which is likely to occur in a warmer atmosphere, containing more moisture. On a regional scale the rise may be counteracted, but also enhanced, by changes in ocean circulation associated with the transition to a warmer climate.

REFERENCES

1. Weber, S. L., H. von Storch, P. Viterbo, and L. Zambresky, Coupling an ocean wave model to an atmospheric general circulation model. *Climate Dynamics,* 9:63–69 (1993).
2. Houghton, J. T., G. J. Jenkins, and J. J. Ephraums, Eds., *Climate Change: The IPCC Scientific Assessment,* WMO/UNEP (Cambridge: Cambridge University Press, 1990).
3. Luther, F. M. and Y. Fouquart, The intercomparison of radiation codes in climate models. *World Climate Programme Report WCP-93* (Geneva: W.M.O., 1984).
4. Land air temperatures from P. D. Jones, Climate Research Unit, Norwich; sea surface temperatures from G. Farmer, Documenting and explaining recent global mean temperature changes (Norwich: Climate Research Unit, Final Report to NERC, UK, contract GR 3/6565).
5. Barnett, T. P. Global sea level change. In: *Climate Variations of the Past Century and the Greenhouse Effect* (Rockville, Maryland: National Climate Program Office/NOAA, 1988).
6. Gornitz, V. and S. Lebedeff, Global sea level changes during the past century. In: Nummedal, D., O. H. Pilkey, and J. D. Howard, Eds. *Sea Level Fluctuations and Coastal Evolution* (Tulsa, Oklahoma: SEPM Special Publication No. 41, 1987) pp. 3–16.
7. Schlesinger, M. E. Equilibrium and transient climatic warming induced by increased atmospheric CO_2. *Climate Dynamics* 1:35–51 (1986).
8. Manabe, S. Transient responses of a coupled ocean-atmosphere model to gradual changes of atmospheric CO_2. Part 1: Annual mean response. *J. Clim.* 4:785–818 (1991).
9. Haarsma, R. J., J. F. B. Mitchell, and C. A. Senior, Tropical disturbances in a GCM. *Climate Dynamics* 8:247–257 (1993).

CHAPTER 2

Recorded Sea Level Variability in the Holocene and Expected Future Changes

Nils-Axel Mörner

ABSTRACT: The Holocene sea level changes are characterized by a general glacial eustatic rise up to about 6000 to 5000 B.P., when the postglacial eustatic rise was completed. After this the global eustatic component was dominated by the redistribution of oceanic water masses, primarily as a function of changes in the main oceanic current systems in response to a feedback coupling with the interchange of angular momentum between the solid earth and the hydrosphere. The long-term sea level trends have continued to operate up to the present, and there are no indications of any drastic changes in this respect. The sea level changes to be expected in the next century are estimated — on the basis of past and present sea level changes and their driving forces — to be on the order of not more than 20 cm, which is significantly less than often claimed.

INTRODUCTION

The Global Change and IGBP projects were formulated for the understanding of the environmental changes to be expected in the near future and for the initiation of a better environmental life-style so that we should try to avoid inducing major future environmental crises and disasters upon planet Earth.

One of the most essential parameters is the sea level change to be expected. In most scenarios, a major future flooding was assumed. Sea level rises in the next century amounting to 1 m, or even more were often claimed. All these high-amplitude figures originate from nonspecialists on sea level changes. More realistic analyses give far less catastrophic values.[1-4] We will analyze the situation; i.e., try to

0-87371-301-X/95/$0.00+$.50
© 1995 by CRC Press, Inc.

learn the lesson from past sea level changes and use this knowledge to interpret present changes and try to predict, or at least estimate, the future sea level changes to be expected.

BACKGROUND PERSPECTIVES

In the 1960s, the debate was both hard and intensive regarding whether sea level rose steadily to the present[5,6] or exhibited a mid-Holocene maximum above present sea level[7] and whether sea level rose smoothly[6,8] or in an oscillatory manner.[7] In 1969, the author was able to demonstrate[9] that in northwestern Europe the sea level, after correcting for local subsidence and uplift, had a low-amplitude oscillatory trend somewhere between the high-amplitude curve of Fairbridge[7] and the smooth curve of Shepard.[6] The author was also able to demonstrate[10] that the "real Holocene sea level problem", in fact, was that practically all so-called "eustatic" curves — though differing significantly in the mid-Holocene — converged at about 8000 B.P. This indicated that the major problem was "an unknown factor" that caused an unequal distribution of the water masses with time.

We now know[11–14] that the search for a global eustatic sea level curve is an illusion, that each region must define its own eustatic sea level changes, and that these changes may differ significantly over the globe and even be of opposite sign. This means that sea level data from different parts of the world must not be mixed together as previously done.[6–8] Nowadays, each region must define its own relative sea level curve from which the crustal components, if possible, later are subtracted so that a regional eustatic curve can be established. Until now this has only been done for northwestern Europe and as a proposition also for the Brazilian coast. Comparisons of different regional eustatic curves and to a lesser degree also regional relative sea level curves provide the means of identifying the compensational redistribution of water masses due to the differential deformation of the dynamic sea surface and/ or the geoid surface.

The factors that cause these differential eustatic changes over the globe are (1) deformations of the geoid relief,[11,15] (2) global isostatic adjustments in response to deglaciation,[16–18] (3) changes in Earth's rate of rotation deforming both the rotational ellipsoid and the dynamic sea surface topography via changes in the ocean circulation system,[12,19,20] and (4) steric changes in the water column by thermal and/or salinity expansion.[4,21]

Figure 2–1 gives an actual satellite profile of the sea surface topography over the western North Atlantic; the geoid surface slopes from +19 m in the northwest to +32 m in the southeast and the dynamic sea surface shows a +4 m rise above the geoid in the center of the Gulf Stream where the water masses are driven together by the actual current forces. It is the dynamic sea surface and/or the geoid level that are deformed with time when we talk about sea level changes or "eustatic" changes.

Whereas we earlier defined "eustasy" as "simultaneous changes in sea level",[7] we must today define "eustasy" simply as "ocean level changes".[12,22]

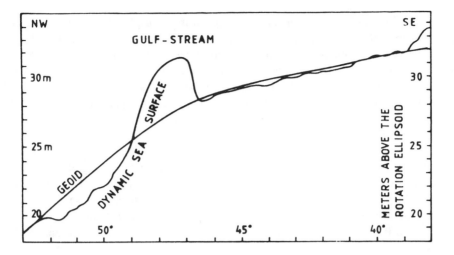

Figure 2–1 Actual satellite profile of the sea surface topography in the western North Atlantic indicating how the strong current forces of the Gulf Stream force the dynamic sea surface to deviate from the geoid level (the geodetic sea level). It is the dynamic sea surface and the geoid level that deform with time when we talk about "eustatic" changes.

Table 2–1 Main Eustatic Sea Level Variables, Their Amplitudes, Maximum Rates, Time Application, and Present-to-Future Significance

Sea level variable	Amplitude	Rate (mm/year)	Period	Significance
Glacial eustasy	100–120 m	10–20	18–5 Ka	Little or none
	>0.5 m	1.0	Last 5 Ka	Some significance
Tectono eustasy	Large	0.06	All time	Insignificant
Global isostasy	20 m	1.0	Last 20 Ka	Insignificant
	5 m	1.0	Last 5 Ka	Insignificant
Geoidal eustasy	180 m relief	5–10	All time	Some significance
	±60 m	5–10	18–10 Ka	—
	±10 m	5–10	Last 8 Ka	—
Rotation ellipsoid	Tens of meters	1.0	All time	Small
Ocean circulation	1–2 m	10	All time	Most significant
Steric effects	>0.5 m	1.0–0.5	All time	Significant

From past sea level records, we learn about the driving forces, causal connections, and the actual amplitudes and rates involved.[12,15,20] Table 2–1 gives the amplitudes and rates established and their possible application today.

THE HOLOCENE MAIN SEA LEVEL TRENDS

The "World Atlas of Holocene Sea-Level Changes"[23] gives a good and comprehensive view of the changes in sea level. Besides global differentiation in the eustatic ocean components, the recorded relative sea level changes are strongly affected by differential tectonics and sedimentary compaction.[12] Despite this complexity, one may distinguish some general regularities. These are illustrated in Figure 2–2.

Five sea level curves from quite different parts of the globe are compared. Despite significantly different shapes, trends, and levels, they all show (1) a general rise up to some 6000 to 5000 B.P. and (2) an out-of-phase to sometimes opposed oscillatory trend for the last 5000 to 6000 years around the present zero level. We interpret the first part as the effect of a general glacial eustatic rise in sea level prior to 6000 to 5000 B.P. (related to the melting of ice caps and glaciers after the last glacial period) plus a local tectonic differentiation (some geoidal differentiation may also apply). The second part (the last 5000 to 6000 years) lacks any significant glacial eustatic component and is instead dominated by the redistribution of water masses around the present mean sea level. This dominant two-phase eustatic sea level trend is illustrated in the basal graph of Figure 2–2; a general glacial eustatic rise up to 6000 to 5000 B.P. is followed by an interregional redistribution of the oceanic water masses. This means that the short-term sea level changes during the last 5000 years primarily are of compensational nature over the globe, i.e., that the volume did not change significantly, but that the water was differentially distributed over the globe due to changes in the transport by the currents and to deformations of the gravity potential surface. Besides this, there might be some steric effect due to warming or cooling of water masses.

The redistribution of water masses during the last 5000 years seems primarily to be the function of changes in the main oceanic current systems in response to the interchange of angular momentum between the "solid" earth and the hydrosphere.[19,20]

THE REDISTRIBUTION OF WATER MASSES

Sea level cannot change without a corresponding response in the earth's rate of rotation.[24] During the deglaciation phase with a general glacial eustatic rise in sea level, Earth experienced a deceleration. When the glacial eustatic rise finished some 6000 to 5000 years ago, the general deceleration finished and was instead overtaken by an interchange of angular momentum between the "solid" earth and the core, the hydrosphere and the atmosphere. The interchange of angular momentum with the hydrosphere — which, according to the present author,[3,19,20] is far more important than previously appreciated — must have generated large-scale changes in the ocean currents in a feedback coupling process. It seems, therefore, quite significant that the sea level changes after some 5000 to 6000 B.P., when the glacial eustatic rise was completed, are dominated by the redistribution of the water masses over the globe.

During the last 300 years, we have instrumental records of the changes in rotation[25,26] and the corresponding retardation/acceleration variations in the major current system[19] leading to the redistribution of water masses (recorded by sea level) and stored energy (recorded by climatic parameters).

Any rise in global sea level must lead to a deceleration in the earth's spin velocity (i.e., an increase in the length of the day, LOD). This physical law was used to investigate the sea level of the last centuries.[3]

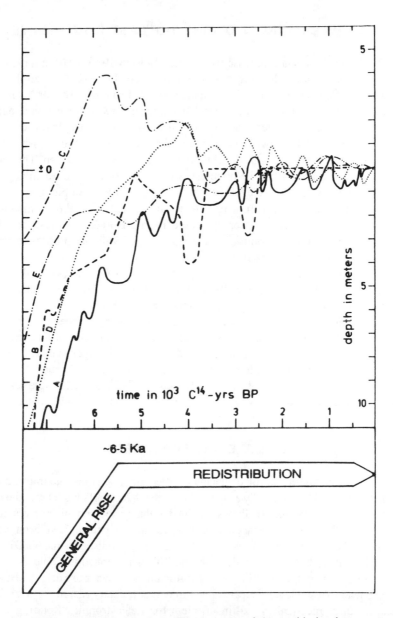

Figure 2–2 Five sea level curves from different parts of the world showing a general rise (despite tectonic differentiation) up to about 6000 to 5000 B.P. and thereafter an out-of-phase (sometimes opposed) variability around a base level close to the present zero, which is interpreted as the effect of irregular redistribution of the oceanic water masses. The two-phase main eustatic trends are illustrated in the lower graph. The five curves represent: (A) the regional eustasy of northwestern Europe,[43] (B) the possible regional eustatic curve of Brazil,[44,45] (C) the relative sea level changes in Vietnam,[46] (D) the relative sea level changes in northern New Zealand,[47,48] and (E) the relative sea level changes in parts of Japan.[49]

THE SEA LEVEL CHANGES DURING THE LAST 100 TO 200 YEARS

In Europe, eustatic sea level rose from about 1830 to 1930–1940 at a mean rate of about 1.1 mm/year.[27] The origin of this rise is not known. A general rise in temperature is often claimed to be the ultimate cause.[28] From eastern North America, a rise of 2.4 mm/year is often advocated.[29] This may perhaps be correct on a regional scale. Such a high rate of sea level rise is impossible, however, on a global scale. This even applies for the analysis by Trumpin and Wahr[30] arriving at a recent rise of about 1.75 mm/year. In other parts of the world, other figures have been given. The picture is far from uniform and rather the opposite is true — it is quite confusing.

Our presently available sea level records give a very divergent picture.[2,31,32] This is not at all surprising — rather what one would expect from the dominance of redistribution of water masses over the globe in late Holocene times[19] and local crustal movements and compaction. Furthermore, the available records primarily represent Northern Hemisphere records.

The best tool for the evaluation of any "global" trend is the analysis of the earth's rotational record (LOD) because any change in the radius has to affect the rotation, and vice versa. Our analysis[3] indicates that (1) either there has been a global rise (1830 to 1930) that amounted to a maximum of 1.1 mm/year, or (2) there has been no general rise at all, merely a cyclic redistribution of water masses. The importance of this analysis is that it shows that there can have been no global sea level rise larger than 1.1 mm/year (which concurs perfectly well with the long-term records of the slowly subsiding Dutch coast) in contrast to statements of much higher rates.[29–30,33-35] It seems significant that Woodworth[36] failed to identify any acceleration in the mean sea level within this century.

CLIMATE AND SEA LEVEL

During the maximum rate of deglaciation after the glaciation maximum around 20,000 B.P., sea level rose glacially eustatically by about 20 to 30 mm/year, at the most. These rates represent extreme conditions confined to this period and this process. In late Holocene times, the rates of glacial eustatic changes must, of course, have been considerably smaller; because the glacial eustatic rise seems to have finished some 6000 to 5000 years ago, it is likely to be very small or absent. This is illustrated in Table 2–2.

The climate may get warmer in the near future due to greenhouse effects. But why should this lead to a sea level rise? Sea level may theoretically rise in the near future by means of three mechanisms, all insufficient for a catastrophic "flooding":

1. *Melting of existing mountain glaciers.* This is not enough because the total water volume in all these glaciers only equals 0.5 m sea level rise,[37] and many of them will remain in existence even with a considerable temperature rise (Table 2–2).
2. *Expansion of the ocean water column.* This can only generate some centimeters to a decimeter rise (the expansion of the surface water can hardly exceed 10 cm, at the very most 20 cm, and the bottom water can only warm after a time lag on the order

Table 2–2 Present Glacial Volume Expressed in Meter Sea Level Equivalent, Their Behavior in Mid-Holocene Times, and Their Prospects for the Near Future

Glacial ice sheets	Volume	Mid-Holocene	Future prospects
Antarctica	60 m	Readvance	Reactions too slow for rapid sea level changes; note readvances at Holocene and Medieval warm optima
Greenland	6 m	Uncertain	Uncertain effects; note major glacial expansion at Thule Base around 1920
Alpine glaciers	0.5 m	Smaller	Limited effects on sea level; at the most 1.0 mm/year

Note: Total glacial eustasy probably does not exceed 1.0 mm/year, which is the long-term effect of the last 5000 years, and is likely to remain so in the near future.

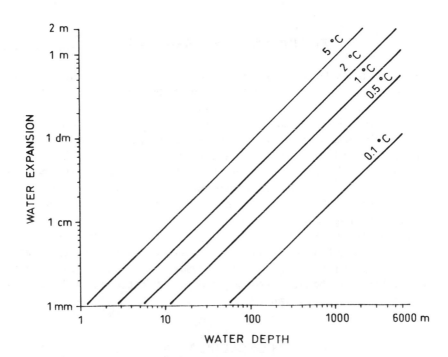

Figure 2–3 Relations among heating (diagonals) and expansion of the ocean water column. Only the surface water is susceptible to rapid changes in temperature. A present global temperature signal will take about 1000 years to reach the bottom water. The reaction times for the intermediate water layers are less well known.

of 1000 years or so). Figure 2–3 gives the relations among warming and expansion of the water column.[4]

3. *Melting of the Antarctic ice cap.* A possible melting is too slow a process to generate any significant effects in the near future (this even applies for a hypothetical West Antarctic surge). During the Holocene climatic optimum with a global mean temperature about 1°C higher than today, global sea level seems not to have been higher. And what is more important, at about 5000 to 6000 B.P. — right during the

mid-Holocene climatic optimum — Antarctica experienced glacier expansions. Similarly, during the Medieval Warm Peak, there was a significant readvance of the ice cap over Livingstone Island on the west side of the South Shetland Islands.[20] This means that warming does not automatically lead to glacier melting in Antarctica; rather the opposite is true due to an increase in precipitation (Table 2–2).

It has been proposed that there has been a close relation between global warming and sea level in the last 100 years.[38–40] This is in no way well documented.[4] First of all, the curves compared do not substantiate the conclusions drawn. This is illustrated in Figure 2–4. Second, the sea level curve used does not represent any global trend. The same can be said about the temperature curve used.

FUTURE SEA LEVEL CHANGES

The future sea level changes to be expected are the function of our understanding of the past and present trends. A present maximum mean sea level rise of 1.1 mm/year fits well with the past long-term records (Table 2–1).

Whether the earth experiences natural cyclic changes[41] or a greenhouse warming[39] needs further clarification. With respect to sea level, for the next 100 years, we can only expect a maximum rise on the order of 10 cm with an additional steric effect of about the same amount, at the most — Nakiboglu and Lambeck[21] calculated the present steric effect at about 0.5 mm/year, which seems quite reasonable (and agrees well with Figure 2–3). This gives a total maximum rise on the order of 20 cm, which is far from a catastrophic "flooding" as sometimes proposed.[33,42] In some areas, even a 20-cm rise would be quite serious, however (especially in combination with local subsidence).

If a greenhouse effect should cause a significant increase in global temperature, this will not lead to any high-amplitude glacial eustatic rise in sea level due to melting glaciers (Table 2–2). Therefore, there are no reasons to expect any high-amplitude glacial eustatic rise due to warming.[20]

CONCLUSIONS

The long-term Holocene sea level trends and causal mechanisms have in no significant way changed in recent times. On the contrary, the governing parameters and forces seem to have continued to control the sea level changes during the last 150 years, and are expected to continue doing so in the near future. This means that there are no reasons to expect any larger sea level rise during the next century — even with a significant warming — of more than 20 cm or so. Catastrophic flooding ideas should be dismissed. This, of course, does not mean that anthropogenic changes of the atmospheric composition is not a very serious problem. In low coastal areas and in combination with local subsidence, even a 10 to 20 cm rise in sea level may give rise to serious problems.

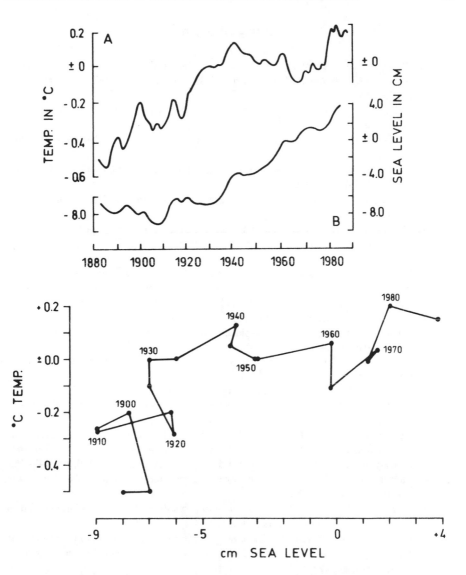

Figure 2–4 Five-year running mean curves of (A) global mean annual temperature according to Hansen and Lebedeff[50] and (B) global sea level changes according to Barnett.[34] These curves have been widely used to demonstrate a close correlation between temperature and sea level.[38] This is not at all the case, however, as demonstrated in the lower curve where the temperature values (A) have been plotted against the sea level values (B) for every 5 years. There is actually little or no straightforward correlation between temperature and sea level. When the temperature rose significantly between 1890 and 1930, sea level hardly rose at all. When the sea level rose significantly between 1930 and 1985, the temperature rise was quite small.

REFERENCES

1. Pirazzoli, P. A. Recent sea-level changes in the North Atlantic, in *Late Quaternary Sea-Level Correlation and Application,* D. B. Scott et al., Eds. (Kluwer Publ., 1989), pp. 153–167.
2. Pirazzoli, P. A. Present and near-future global sea-level changes, *Palaeogeogr. Palaeoclim. Palaeoecol.,* 75:241–258 (1989).
3. Mörner, N.-A. Sea level changes and Earth's rate of rotation, *J. Coastal Res.,* 8:966–971 (1992).
4. Mörner, N.-A. Global sea level in the past and future century, *Bull. INQUA Neotrectonics Comm.,* 16:64–66 (1993).
5. Jelgersma, S. Holocene sea-level changes in the Netherlands, *Meded. Geol. Sticht.,* Ser. C, Vi (7):1–101 (1961).
6. Shepard, F. P. Thirty-five thousand years of sea level, in: *Essays in Marine Geology in Honour of K. O. Emery,* T. Clements, Ed. (Univ. S. California Press, 1963), pp. 1–10.
7. Fairbridge, R. W. Eustatic changes in sea level, *Phys. Chem. Earth,* 4:99–185 (1961).
8. Bloom, A. L. Sea-level movements during the last deglacial hemicycle. Geological correlations, Report of the International Correlation Program (IGCP) (1983) 11:22.
9. Mörner, N.-A. The Late Quaternary history of the Kattegatt Sea and the Swedish West Coast; deglaciation, shorelevel displacement, chronology, isostasy and eustasy, *Swed. Geol. Survey,* C-640:1–487 (1969).
10. Mörner, N.-A. The Holocene eustatic sea level problem, *Geol. Mijnbouw,* 50:699–702 (1971).
11. Mörner, N.-A. Eustasy and geoid changes, *J. Geol.,* 84:123–152 (1976).
12. Mörner, N.-A. Models of global sea level changes, in *Sea Level Changes,* M. J. Tooley and I. Shennan, Eds. (Blackwell, 1986), pp. 333–355.
13. Newman, W. S., L. Marcus, R. Pardi, J. Paccione, and S. Tomacek, Eustasy and deformation of the geoid: 1000–6000 radiocarbon years BP, in: *Earth Rheology, Isostasy and Eustasy,* N.-A. Mörner, Ed. (John Wiley & Sons, 1980), pp. 449–463.
14. Pirazzoli, P. A. Global sea-level changes and their measurements, *Global Planetary Change,* 8:135–148 (1993).
15. Mörner, N.-A. Sea levels, in: *Mega-Geomorphology,* R. A. M. Gardner and H. Scoging, Eds. (Oxford University Press, 1983), pp. 73–91.
16. Clark, J. A. A numerical model of worldwide sea-level changes on a viscoelastic Earth, in: *Earth Rheology, Isostasy and Eustasy,* N.-A. Mörner, Ed. (John Wiley & Sons, 1980), pp. 525–534.
17. Clark, J. A., W. E. Farrell, and W. R. Peltier, Global changes in postglacial sea level: a numerical model, *Quat. Res.,* 9:265–287 (1978).
18. Peltier, W. R. Dynamics of the ice age Earth, *Adv. Geophys.,* 24:1–144 (1982).
19. Mörner, N.-A. Terrestrial variations within given energy, mass and momentum budgets; Paleoclimate, sea level, paleomagnetism, differential rotation and geodynamics, in: *Secular Solar and Geomagnetic Variations in the Last 10,000 Years,* F. R. Stephenson and A. W. Wolfendale, Eds. (Kluwer Acad. Press, 1988), pp. 455–478.
20. Mörner, N.-A. Global change. The last mellennia, *Global Planetary Change,* 7:211–217 (1993).

21. Nakiboglu, S. M. and K. Lambeck, Secular sea-level changes, in: *Glacial Isostasy, Sea-Level and Mantle Rheology,* R. Sabadini, K. Lambeck, and E. Boschi, Eds. (Kluwer Acad. Publ. Press, 1991), NATO C-334, pp. 237–258.
22. Mörner, N.-A. The concept of eustasy. A redefinition, *J. Coastal Res.,* SI-1:49–51 (1986).
23. Pirazzoli, P. A. World atlas of Holocene sea-level changes, *Elsevier Oceanogr. Ser.,* 58:1–300 (1991).
24. Dieke, R. H. The secular acceleration of Earth's rotation and cosmology, in: *The Earth-Moon System,* B. G. Marsden and A. G. W. Cameron, Eds. (Plenum Press, 1966), pp. 94–164.
25. Rochester, M. G. Causes of fluctuations in the rotation of the Earth, *Phil. Trans. R. Soc. Lond.,* A 313:95–105 (1984).
26. Stephenson, F. R. The Earth's rotation as documented by historical data, in: *New Approaches in Geomagnetism and the Earth's Rotation,* S. Flodmark, Ed. (World Sci. Publ., Singapore, 1991), pp. 87–113.
27. Mörner, N.-A. Eustatic changes during the last 300 years, *Palaeogeogr. Palaeoclim. Palaeoecol.,* 13:1–14 (1973).
28. Fairbridge, R. W. and A. O. Krebs, Jr., Sea level and southern oscillation, *Geophys. J. R. Astr. Soc.,* 6:532–545 (1962).
29. Peltier, W. R. and A. M. Tushinghan, Global sea level rise and the greenhouse effect: might they be connected? *Science,* 244:806–810 (1989).
30. Trumpin, A. S. and J. M. Wahr, Constraints on long-period sea level variations, in: *Glacial Isostasy, Sea-Level and Mantle Rheology,* R. Sabadini, K. Lambeck, and E. Boachi, Eds. (Kluwer Acad. Publ. Press, 1991), NATO C-334, pp. 271–284.
31. Pirazzoli, P. A. Secular trends of relative sea-level (RSL) changes indicated by tide-gauge records, *J. Coastal Res.,* SI-1:1–26 (1986).
32. Pirazzoli, P. A., D. R. Grant, and P. Woodworth, Trends of relative sea-level changes: past, present, future, *Quat. Int.,* 2:63–71 (1989).
33. Hoffman, J. S., D. Keyes, and J. G. Titus, *Projecting Future Sea Level Rise,* (Environmental Protection Agency, U.S. Government Printing Office, Washington, D.C., 1983).
34. Barnett, T. P. Global sea level changes, in: *Climate Variations over the Past Century and the Greenhouse Effect,* (National Climate Program Office Report. NCPO/NOAA, Rockville, MD, 1988).
35. *Global Change: Reducing Uncertainties* (IGBP, 1992).
36. Woodworth, P. L. A search for accelerations in records of European mean sea level, *Int. J. Climatol.,* 10:129–143 (1991).
37. Flint, R. F. *Glacial and Quaternary Geology* (Wiley, 1971).
38. IGBP Report 12 (1990).
39. World Climate Research Program, *Global Climate Change* (WMO & ICSU, 1990).
40. Visconti, G. Global warming expected from increase of greenhouse gases: a forcing for sea level changes, in: *Glacial Isostasy, Sea-Level and Mantle Rheology,* R. Sabadini, K. Lambeck, and E. Boschi, Eds. (Kluwer Acad. Publ. Press, 1991) NATO C-334, pp. 203–212.
41. Friis-Christensen, E. and K. Lassen, Length of the Solar cycle: an indicator of Solar activity closely associated with climate, *Science,* 254:698–700 (1991).
42. Kaplin, P. A. Shoreline Evolution during the Twentieth Century (UNAM Press, Mexico, 1989).

43. Mörner, N.-A. The northwest European "sea-level laboratory" and regional Holocene eustasy, *Palaeogeogr. Paalaeoclim. Palaeoecol.,* 29:281–300 (1980).
44. Martin, L., J.-M. Flexor, D. Blitzkow, and K. Sugio, Geoid change indications along the Brazilian coast during the last 7000 years, *Proc. 5th Int. Coral Reef Congr.* (Tahiti), 3:83–90 (1985).
45. Mörner, N.-A. Eustasy, paleogeodesy and glacial volume changes, in: *Sea Level, Ice and Climatic Changes* (IAHS Publ., 1981), pp. 277–280.
46. Fontaine, H. and G. Delibrias, Niveaux marins pendant le Quaternaire au Viet-Nam, *Arch. Géol. Viet-Nam,* 17:35–44 (1974).
47. Suggate, R. P. Post-glacial sea-level rise in the Christchurch Metropolitan area, New Zealand, *Geol. Mijnbouw,* 47:291–297 (1968).
48. Schofield, J. C. Sea level fluctuations during the last 4000 years as recorded by chenier plain, Firth of Thames, New Zealand, *N.Z. J. Geol. Geophys.,* 18:295–316 (1960).
49. Maeda, Y. Holocene transgression in Osaka Bay and Harima Nada, *Umi to Sora,* 56:145–150 (1980).
50. Hansen, J. E. and S. Lebedeff, Global trends of measured surface air temperature, *J. Geophys. Res.,* 92:13345–13372 (1987).

CHAPTER 3

Present and Future Sea Level: The Effects of Predicted Global Changes

Eric C. F. Bird

SEA LEVEL

The level of the sea relative to the land, around the world's coastlines is determined primarily by the volume of water in the oceans (which depends on the balance of the hydrological cycle) and the size and shape of the crustal depressions occupied by sea water. As has been indicated in the previous chapter (Jelgersma and Tooley), there have been fluctuations in sea level through geological time, although during the past five to six thousand years there has been relative stability, except on coasts where the land has been rising or subsiding. This chapter will deal with contemporary changes in sea level, the forecast of a global rise of sea level, and the effects that such a rise will have on the world's coastlines.

Statements about present sea level usually refer to mean sea level, approximated by the long-term average of high- and low-tide levels, i.e., mean tide level. In practice, this is defined in relation to some national datum such as American Sea Level Datum, or Ordnance Datum in the United Kingdom.[1] Globally, tide ranges vary from zero to more than 20 m; there are cyclic variations from neap to spring tides, and longer-term oscillations up to the 18.6-year cycle, all of which have to be excluded by averaging to determine present sea level. Detection of any change in sea level at a coastal station therefore requires records spanning at least 20 years, to avoid being misled by changes due to long-term tidal oscillations. In addition there are fluctuations of sea level related to meteorological conditions, notably temperature and pressure changes, as well as short-term variations related to the passage of cyclonic and anticyclonic pressure systems with their associated winds, which can produce storm surges that raise sea level temporarily by up to several meters. There are also long term, often seasonal variations related to climatic patterns, notably

0-87371-301-X/95/$0.00+$.50

temperature and wind regimes. Global climatic phenomena such as the El Niño Southern Oscillation can also modify sea level along sectors of coastline.[2] These variations, which complicate the oscillations registered on tide gauges, must also be excluded from the record when present sea level is determined.

When allowance is made for all these variations it is found that most tide gauge records show evidence of a rising, falling, or fluctuating sea level. Among the longest continuous records are those from San Fransisco (U.S.A.), Brest (France), and Sydney (Australia), each of which shows a generally upward movement. Several attempts have been made to determine global trends in sea level from statistical analyses of long-term tide gauge records,[3] and evidence from repeated geodetic surveys. There is wide support for the idea that global sea level has risen 10 to 15 cm during the past century, and that it is now rising by about 1.2 mm/year.[4]

However, as Pirazzoli[5] demonstrated, tide gauge records from around the world over recent decades have shown variations in mean sea level trends. Of 229 tide gauge stations with at least 30 years of reliable recent records, 63 (28.5%) showed a mean sea level rise >2 mm/year, 52 (22.5%) 1 to 2 mm/year, and 47 (20.6%) <1 mm/year, the remaining 65 (28.5%) having shown a mean sea level fall. As over 70% of these records showed a positive trend it seems probable that there has been a global sea level rise, but such a conclusion may be premature in view of the uneven distribution of the 229 stations, only 6 of which are in the southern hemisphere. It is expected that the Global Sea Level Observing System (GLOSS) recently established by UNESCO, with a much more representative network of tidal stations around the world's coastline, will be able to provide more accurate information in the next few decades. Meanwhile, the suggestion that mean sea level is rising globally should be treated with caution, bearing in mind the geographical variability of Holocene fluctuations of sea level.

Sea level changes over the past few decades should be considered against the background of factors known to have influenced Holocene sea level trends. The most obvious of these are tectonic movements, upward or downward, of coastal land margins. Sea level has been falling on coasts where tectonic land uplift has been in progress, as in the regions of Scandinavia, northern Canada, and Alaska where isostatic uplift has followed late Quaternary deglaciation. In northeastern Sweden and northern Finland, for example, tide gauge records have shown uplift proceeding at up to 1 cm/year, confirmed by evidence from repeated geodetic surveys, and substantial areas of former sea floor have emerged as new coastal lowland in recent centuries. These regions have been excluded from the global analyses of tide gauge records mentioned previously. Emerging coastal sectors are also to be found in tectonically active regions such as northern New Guinea and parts of Indonesia and the Philippines.

There has also been a growing awareness of the extent of subsiding coasts in recent years (Figure 3-1). Tectonic subsidence, gradual or intermittent, has contributed to many of the records of mean sea level rise >2 mm/year determined by Pirazzoli,[5] which include deltaic regions where accumulating fluvial sediment loads have been downwarping the earth's crust.

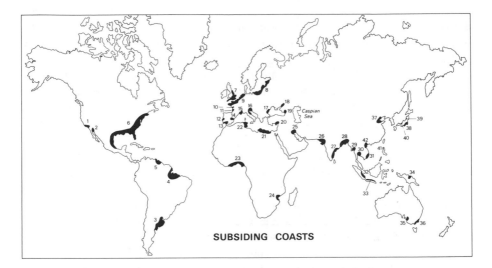

Figure 3–1 Subsiding coasts, where the relative rise of sea level indicated by tide gauge records has averaged more than 2.0 mm/year over the past century, and where there is evidence of continuing subsidence. These coasts show submergence and erosion which will become much more widespread around the world's coastline if there is a global sea level rise. Key: 1, Long Beach area, southern California; 2, Columbia River Delta, head of Gulf of California; 3, Gulf of La Plata, Argentina; 4, Amazon Delta; 5, Orinoco Delta; 6, Gulf/Atlantic coast, Mexico and United States; 7, southern and eastern England; 8, The southern Baltic from Estonia to Poland; 9, northern Germany, the Netherlands, Belgium and northern France; 10, Loire Estuary, western France; 11, Vendée, western France; 12, Lisbon region, Portugal; 13, Guadalquavir Delta, Spain; 14, Ebro Delta, Spain; 15, Rhône Delta, France; 16, northern Adriatic from Rimini to Venice and Grado; 17, Danube Delta, Rumania; 18, eastern Sea of Azov; 19, Poti Swamp, Soviet Black Sea coast; 20, southeastern Turkey; 21, Nile Delta to Libya; 22, Tunisia; 23, Nigerian coast; 24, Zambezi Delta; 25, Tigris-Euphrates Delta; 26, Rann of Kutch; 27, eastern India; 28, Ganges-Brahamputra Delta; 29, Irrawaddy Delta; 30, Bangkok coastal region; 31, Mekong Delta; 32, eastern Sumatra; 33, northern Java Deltaic coast; 34, Sepik Delta; 35, Port Adelaide region; 36, Corner Inlet region; 37, Hwang-Ho Delta; 38, Tokyo; 39, Niigata, Japan; 40, Maizuru, Japan; 41, northern Taiwan; 42, Red River Delta, North Vietnam. The Caspian Sea has risen more than 1.5 m since 1977 (see Figure 3-3).

Tectonic movements that result in changes in the shape and size of the ocean basins will also lead to global sea level changes, which have been styled tectono-eustatic to distinguish them from the downward and upward movements of sea level that accompany the waxing and waning of the world's ice sheets, glaciers, and snowfields: so-called glacio-eustatic. In addition, it has been realized that a sea level rise can be augmented by hydro-isostatic subsidence, especially on low-lying coasts where wide continental shelves have been depressed beneath the weight of the deepening sea, especially where soft compressible sediments, such as peat, occur beneath the sea floor.

Human activities such as groundwater extraction and depletion of aquifers under coastal urban centers, such as Venice and Bangkok, or oil extraction on the parts of the coast of southern California have caused localized coastal subsidence. Figure 3-1 shows the global distribution of coasts known to have been subsiding in recent decades, and it is probable that further research will increase the number and extent of these.

Contemporary sea level changes are also likely to be influenced by variations in the ocean surface resulting from continuing changes in geodetic sea level, Mörner[6] pointed out that the surface of the oceans is complicated by the presence of high and low areas with a relief of the order of 180 m, and that these are partly related to relatively stable gravitational patterns and partly to more the dynamic forces that produce circulations of ocean water. Their distribution has undoubtedly changed in the past, and any continuing change will be marked by a sea level rise where high areas move coastward and a sea level fall where low areas move in. Such changes could augment, negate, or outweigh fluctuations due to other causes.

Other factors that will modify sea level changes include the inflow of sediment, notably from rivers and melting glaciers, to build up the sea floor, and the impacts of human constructions such as ports and land reclamation schemes, which have changed coastal configurations and modified local tide regimes.[7] The extent of a sea level change may also vary with modifications of tidal amplitude related to changes in coastal and nearshore configuration: for example, where tide range now diminishes into embayments with narrow entrances (e.g., Port Phillip Bay, Australia) a sea level rise is likely to increase them, and where they are presently amplified towards the heads of funnel-shaped inlets (e.g., Bay of Fundy, Canada) a sea level rise may reduce them.

Finally, it should be borne in mind that tide gauge records come mainly from instruments that were installed for the purposes of harbor procedures and ship navigation. Many are located on port structures where there are local tidal anomalies resulting from wave reflection and surge ponding, and may also be exposed to damage and disturbance in the course of ship movements and port operations. Any subsidence of their foundations will of course be registered as a rise in mean sea level. The GLOSS network will ensure more scientific instrumentation, and more reliable monitoring of sea level changes, supported by increasingly accurate information from repeated geodetic surveys, especially with the use of data from satellites.

BIOLOGICAL ZONATIONS AS SEA LEVEL INDICATORS

The limitations of tide gauge records and geodetic surveys as indications of sea level changes have prompted consideration of biological zonations as sea level indicators.[1] Evidence of sea level changes may be obtained from repeated surveys of the levels of marine organisms such as oysters, mussels, algae, and kelp, that are often found in vertical zonations, encrusting cliffs, rocky shores, sea walls, and pier supports. These levels are correlated with the depth and duration of marine submergence,[8] and can be expected to move in relation to a rise or fall of sea level. They should move upward on cliff faces, stacks, and rocky protrusions, as well as artificial structures, as sea level rises, and landward across shore platforms and intertidal outcrops.

Evidence that they can do so has been reported from the subsiding coast of southeastern Florida by Wanless,[9] who showed that horizons of oysters and barnacles moved 15 cm upwards on concrete pilings at Miami Beach between 1949 and 1981, consistent with a rise of mean sea level registered on nearby tide gauges. It may be possible to detect such migrations with reference to historical photographs where mean sea levels can be determined in relation to fixed features such as steps or decking, and it would be useful to document existing levels of zoned organisms as a basis for future measurements. In southeastern Australia the calcareous tubeworm *Galeolaria caespitosa* occupies a well-defined intertidal horizon, but within Port Phillip Bay, where tide gauges show little evidence of any recent sea level change, there is no definite indication of any upward movement.[10]

THE GREENHOUSE EFFECT AND SEA LEVEL RISE

It is against this background that we can consider the prospect of a global sea level rise, produced by the augmented greenhouse effect and global atmospheric warming. Measurements initiated during the International Geophysical Year (1957) have shown that the atmospheric concentrations of greenhouse gases (notably carbon dioxide, methane, and nitrous oxide) produced by industrial and agricultural activities have been increasing in the earth's atmosphere. Existing temperatures on the earth's surface are maintained by the greenhouse effect, whereby the atmosphere intercepts some reflected solar radiation and thus retains heat. The increasing greenhouse gases will enhance this effect, making the earth's atmosphere more opaque to reflected radiation, so that still more of the solar energy received at the Earth's surface is retained.

It has been predicted that the greenhouse gases will double during the coming century, resulting in an increase of 1.5 to 4.5°C in the mean temperature of the lower atmosphere. Such an increase will cause an expansion of the volume of near-surface ocean water (the steric effect), and partial melting of the world's snowfields, ice sheets, and glaciers, releasing water into the oceans. The outcome will be a world-wide sea level rise.[11]

There have been various estimates of the scale of this sea level rise. The most recent predictions, based on analyses presented to the Intergovernmental Panel on Climatic Change in 1990, are that global mean sea level will rise 20 cm by the year 2030, and about 60 cm by the end of the next century, attaining 1 m sometime in the 21st century (Figure 3-2). Even if concerted international action halted the discharge of greenhouse gases into the atmosphere by the year 2030, global sea level would still continue to rise, attaining about 50 cm by the year 2100, and proceeding for many more decades before it stabilized.[12] Eventually, global atmospheric management could restore the conditions which existed during the 20th century, and return the sea to its present level. Meanwhile, it is necessary to consider the possible effects of a global sea level rise of 1 m over the next 100 to 200 years, and to refine these as more accurate predictions come to hand.

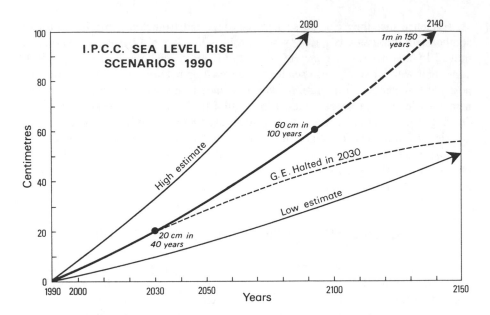

Figure 3–2 Scenarios of sea level rise prepared by the Intergovernmental Panel on Climatic Change. The most likely scenario is a rise of 20 cm by the year 2030, 60 cm by 2090, and 1 m by 2140, but high estimates (1 m by 2090) and low estimates (50 cm by 2150) were also offered. Halting of the enhanced greenhouse effect by the year 2030 would give a decelerating sea level rise, indicated by the broken line, with eventual stability after a rise of more than 50 cm.

Useful evidence may be expected from subsiding coasts (Figure 3-1), where the geomorphological and ecological effects of a sea level rise can already be studied. The effects of submergence can also be seen on the shores of the Caspian Sea, the level of which has risen more than 1.5 m since 1977, a far more rapid transgression than is predicted globally (Figure 3-3).[13]

GENERAL EFFECTS OF A RISING SEA LEVEL

Most existing coastal features have developed during the period of relatively stable sea level that prevailed as a sequel to the worldwide late Quaternary marine transgression, which began about 18,000 years ago, and brought the sea up to approximately its present level about 6000 years ago (see previous chapter). Coasts where the sea level stillstand has endured for up to 6000 years have developed such features as broad shore platforms fronting cliffed coasts, wide coastal plains with multiple beach or dune ridges, prograded deltas and coastal plains, and extensive intertidal zones partly occupied by salt marshes and mangroves.

Coastlines will continue to change in response to coastal processes, notably wave and wind action, tidal and other currents, and weathering. Changes particularly

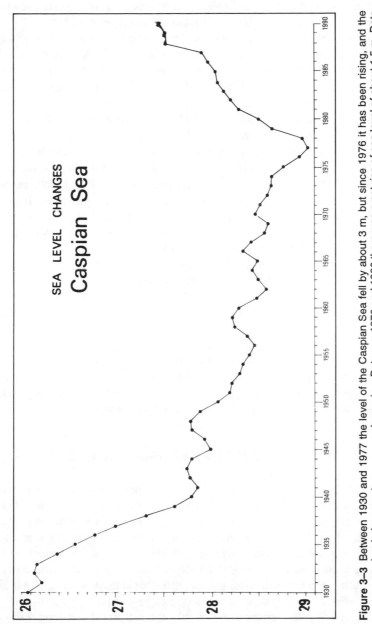

Figure 3–3 Between 1930 and 1977 the level of the Caspian Sea fell by about 3 m, but since 1976 it has been rising, and the bordering coasts are now submerging. Between 1970 and 1990 there was a net rise of sea level of about 1.5 m. Data supplied by the late Professor O. K. Leontyev, and by Dr. S. A. Lukyanova, Moscow State University.

noticed are those which occur during extreme events, such as storm surges, when dramatic erosion and deposition take place. The monitoring of coastline changes is required to determine the nature and rates at which such changes are already taking place, as a basis for assessing what will happen as sea level rises.[14]

If global sea level rises 1 m in the manner predicted, there will be extensive marine submergence of low-lying coastal areas. The highest tides will reach at least a 1 m above their present limits, for there will generally be a slight increase in coastal tide ranges as the oceans deepen. The low-tide line will also move landward, so that at least part of the existing intertidal area will become permanently submerged. In detail, there will be variations related to coastal configuration and nearshore sea floor morphology, and accompanying erosion and deposition.

On hard rock coasts the high-tide line after a 1 m sea level rise can be estimated by surveying the contour at this level above present high tide, but on most other coasts the extent of submergence will be increased by erosion as nearshore waters deepen, so that larger and more destructive waves break upon the shore. The extent to which the coastline will retreat as the result of a 1 m sea level rise will also be modified where there is accompanying uplift or depression of the land margin.

The most obvious effect of a sea level rise will be to initiate erosion on coasts that are at present stable or accreting, and accelerate it where it is already receding. Erosion will further increase where climatic changes lead to more frequent and severe storms, and surges that penetrate farther than they do now. The extent of erosion will also depend on how the nearshore sea floor is modified by the rising sea as the coastline moves landward. Although there will generally be deepening of nearshore waters and a consequent increase in wave energy approaching the coastline, there will be some sectors where an increase in sediment supply derived from fluvial or alongshore sources maintains, or even shallows, the nearshore profile as the sea rises. On such coasts wave energy will not be augmented, and there may be little, if any, coastline erosion; there may even be some progradation.

OTHER EFFECTS OF GLOBAL WARMING

As global warming proceeds, some regions are expected to receive more rainfall while others become drier.[12] Some rivers will carry more sediment because of heavier catchment runoff; many already have increased loads because of the depletion of catchment vegetation by increasing aridity, or by clearance, especially where forest areas have been adapted for agricultural use. Additional coastal sediment could be derived from more rapidly eroding cliffs. It is possible that on some coasts an increasing sediment supply will result in deposition that at least partly offsets the land losses that would otherwise occur as submergence proceeds.

As well as raising temperatures in coastal regions, global warming will cause migration of climatic zones, tropical sectors expanding poleward as the domain of arctic coasts shrinks. Some coastal regions will become wetter, others drier. Tropical cyclones will extend into higher latitudes, bringing storm surges and torrential downpours to coasts that now lie outside their range. Coasts now subject to recurrent storm surges will have still more extensive marine flooding as submergence proceeds, and more severe erosion and structural damage where larger waves reach the coast. Coastal regions where rainfall increases will have more frequent and persistent river flooding than they do now, and the water table rise will be added to that caused by the rising sea level. Some low-lying parts of coastal plains will become permanent swamps or lakes, the salinity of which will depend on the interaction between increasing marine incursion (and perhaps an upward movement of subterranean salt) and the offsetting effects of augmented rainfall and freshwater runoff. Coastal regions that become drier will have more extensive marine salinity penetration into both surface and underground water.

Most marine plants have specific latitudinal ranges, depending on sea temperature, which will increase as global warming proceeds. Thus mangroves, now largely confined to tropical coasts, will extend their range poleward along coasts that provide suitable habitats, whereas the distribution of kelp, a plant restricted to cooler waters, will contract. Many temperate salt marsh species will also move poleward, and it is probable that corals, like mangroves, will extend northward and southward beyond their present latitudinal limits.

PREDICTIONS

Because of the complexity of coastal systems, and their variations from sector to sector, a great deal of site-specific research will be needed to predict the extent of the geomorphological and ecological changes that will occur as sea level rises over the coming century.[15] It must be realized that coastal landforms and ecosystems have been subjected to many changes, both natural and due to human activities, in recent centuries, that changes are still in progress, and that they will continue even if the sea were to remain at its present level. It becomes a question of identifying the changes that will occur as a result of sea level rise, in addition to those that would have taken place anyway.[16] Attempts to predict the extent of coastal changes as sea level rises have run into a number of difficulties because knowledge of coastal hydrodynamics, sediment flow patterns, and their interactions has proved inadequate for generating predictive models.

In assessing the nature and extent of such changes, it is useful to classify coasts into several categories: steep and cliffed coasts, beach-fringed coasts, estuaries and coastal lagoons, deltaic coasts, salt marshes and mangrove coasts, intertidal and nearshore areas, coral reefs, and artificial (man-made) coasts.

STEEP AND CLIFFED COASTS

Cliffs have formed on coasts where the late Quaternary marine transgression brought the sea alongside high ground, with wave action sufficiently strong to be erosive. They are generally fronted by shore platforms, which typically decline seaward from high-tide level at the cliff base out to below low-tide level, but on some coasts are almost horizontal. Elsewhere there is either a rugged rocky shore topography, or a plunging cliff descending to deep water inshore.

Cliff morphology is related to geological structure, rock type, tide range, and exposure to wave attack. The present development of cliffs and shore platforms is largely a consequence of the long phase of relatively stable sea level around much of the world's coastline during the past 6000 years. There has been little cliffing on coasts that have been emerging, as in the northern Baltic region, where the subsiding zones of weathering and erosion have withdrawn seaward across an emerging sea floor.

Existing rates of cliff retreat vary with such factors as rock resistance, structure, the presence or absence of shore platforms, exposure to wave action, and tide range. Where soft materials, such as volcanic ash or glacial drift, are exposed to strong wave attack, cliffs have been retreating hundreds of meters in a year; at the other extreme, cliffs in hard rocks such as massive granite have shown little change over many centuries.[14]

Other steep coasts consist of bluffs, mantled with weathered material and soils, held in place by a scrub or forest cover, rather than cliffs that expose rocky outcrops and are actively receding. Some of these bluffs originated as subaerially weathered slopes in Pleistocene times, and have persisted during and since the late Quaternary marine transgression either because the sea that rose alongside them was relatively calm, or because there was some protective feature, such as a coastal terrace or wide beach, which prevented them being undercut by wave action. There are many such steep forested coastal slopes in the humid tropics, notably in the Indonesian Archipelago,[17] where there is occasional slumping of the weathered mantle, especially after the slope foot has been undercut by storm waves.

A rising sea level will generally deepen nearshore waters and submerge at least part of the existing intertidal zone, thereby intensifying wave attack at the base of cliffs and bluffs, and accelerating erosion.[18] Existing cliffs will become more unstable, and will recede more rapidly as sea level rises. Bluffs will be undercut to form basal cliffs, and slumping will become more frequent on the vegetated slopes, which will eventually become steep retreating cliffs. The exception will be where rock outcrops are so resistant that the high- and low-tide lines simply move up the existing cliff face.

Generally, cliff recession has been intermittent, cliff crests receding during episodes of storm erosion or slumping, then remaining in position, perhaps for many decades. A rising sea level will increase the frequency of these events, but it will be difficult to recognize an acceleration of cliff retreat, at least in the early stages, except where existing average rates of recession have been measured with sufficient accuracy to provide a basis for comparison.[14]

On less resistant rocks, such as soft sandstones or clays, waves have shaped the nearshore sea floor into a concave profile that declines seaward from the cliff base, flattening below low tide. As cliffs recede, the nearshore sea floor tends to be lowered in such a way as to move this concave profile landward. This has been the case, for example, along the rapidly receding Holderness coast of eastern England, where cliffs and sea floor are cut into Pleistocene glacial drift deposits. A sea level rise is likely to produce further landward migration of concave sea floor profiles as the cliff base retreats.

Where cliffs of soft material have developed on coasts sheltered from strong wave energy they may be dominated by gulleying, slumping, and other features resulting from subaerial weathering and the effects of runoff and seepage, rather than from marine processes.[19] On such cliffs, a rising sea level is likely to increase marine erosion, reducing and eventually suppressing the features developed by subaerial processes, as undercutting of the cliff base produces a steeper or vertical cliff.

Rates of cliff recession are also influenced by the availability of rocky debris, including beach material, which can be mobilized by wave action and used as ammunition for cliff-base abrasion. Cliffs fringed by a narrow beach that can be used in this way have retreated more rapidly than cliffs on the same formation where the beach is wide and high (so that waves do not reach the cliff base), or where there is no beach material (so that the cliff base is attacked only by the hydraulic action of breaking waves). If a rising sea submerges or disperses a narrow beach in such circumstances, the effects of such abrasion will diminish and, with purely hydraulic action, cliff retreat could decelerate.

Where shore platforms have developed as intertidal features in front of cliffs or bluffs they will be submerged for longer periods, and permanently inundated where the sea level rise exceeds the tidal range: waves that now reach the base of the cliff only at high tide will be able to maintain a more consistent attack, thereby accelerating cliff erosion. Submerging shore platforms may acquire a veneer of sediment, or be mantled by accretionary growths of nearshore plant and animal communities, in which case wave attack will be somewhat diminished. If deposition produces a persistent talus apron or a prograded beach in front of the cliff as sea level rises, wave action will be impeded and cliff retreat will slow down, and perhaps come to a halt.

On some cliffed coasts a basal notch has been formed close to high-tide level by abrasion or solution. There are well-known examples on the limestone stacks of Phangna Bay in Thailand, and similar notches have been found at various levels above and below the present, notably on Pacific islands and around the Mediterranean, where Pirazzoli has correlated them with earlier stillstands of sea level.[20] A rising sea will enlarge existing notches upwards, rather than develop a new notch at a higher level; the latter would form only when a rapid sea level rise was followed by a new stillstand.

BEACH-FRINGED COASTS

Beaches, generally of sand or shingle (beach gravel), are found where sediment has been supplied to the coast by rivers, derived from the erosion of cliffs and rocky shores, or washed in from the sea floor. Some sandy beaches have been supplied with wind-blown sediment from dunes spilling from the hinterland, and in recent decades there has been artificial nourishment of selected beaches by the dumping of sand, gravel, or other suitable rocky debris.[21]

Many beaches are narrow, fringing cliffed coasts or the shores of deltas or coastal plains, but on some coasts beach material has accumulated in such a way as to form beach-ridge plains or coastal barriers and barrier islands in front of lagoons and swamps.[22] Some barriers are anchored in position, and have prograded intermittently by the addition of successively built beach and dune ridges, as on the Gippsland coast in southeastern Australia;[23] others are transgressive, moving landward because of repeated storm overwash and the drifting of dunes driven by onshore winds, as on parts of the Gulf and Atlantic coasts of the United States.[24] In general, anchored and prograded barriers are found on stable or emerging coasts, whereas transgressive barriers are best developed on coasts that are subsiding.

Most beaches show alternations, known as "cut and fill", with erosion when some of their sediment is withdrawn seaward during episodes of storm wave activity, and deposition in intervening periods when gentler wave action moves sediment shoreward to be deposited on the beach face. A beach system may be said to be in equilibrium when the transverse profile of the beach and adjacent nearshore zone alternates cyclically, in response to processes generated by wave action in nearshore waters, without long-term (>1 year) changes — an equilibrium that is lost on beaches that are either advancing by deposition (prograding) or retreating through erosion.

A worldwide study of coastline changes by the International Geographical Union's Commission on the Coastal Environment between 1972 and 1984 demonstrated a global prevalence of beach erosion. It found that in recent decades there has been erosion on more than 70% of the world's sandy coastlines, less than 10% having prograded, with the remaining 20 to 30% having either remained stable or been subjected to alternations with no net change.[25]

There are several reasons for the onset of erosion on beaches that were previously stable or prograding. Beach erosion has generally resulted from a combination of several factors, one or two of which may have been dominant. On some coasts it is mainly because the supply of sediment from rivers has diminished because of dam construction and sediment interception upstream by reservoirs. On others there has been a reduction in sand supply from alongshore (especially where cliffs have been stabilized or intercepting breakwaters built), or from the sea floor. Some beaches have been quarried for sand and gravel, and also for minerals such as rutile, tin, or gold. There are places where wave attack has increased because of the deepening of nearshore water, either because of dredging or as the result of subsidence.

There is no doubt that a global sea level rise will initiate beach erosion, or accelerate it where it is already taking place, but it has not been the primary cause of the widespread beach erosion that has already occurred during the past century. Beaches have been eroding even on parts of the Baltic coast where the land has risen relative to the sea, as on the Indalsälven Delta (Sweden), the erosion being primarily due to a recent reduction in fluvial sediment yield following dam construction upriver.[14] Nevertheless, a global sea level rise will undoubtedly result in beach erosion becoming even more extensive and severe than it is now.

As sea level rises, submergence will generally result in a deepening of nearshore water, so that larger waves break upon the shore. In most cases beach erosion will increase. Beaches will disappear from sectors where they are already narrow, and

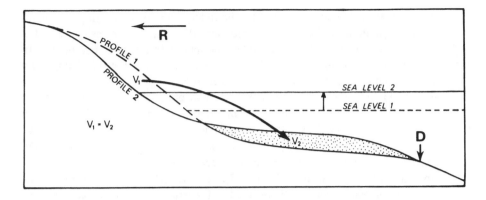

Figure 3-4 The Bruun Rule states that a sea level rise will lead to retreat of the coastline (R) due to erosion of the beach and deposition nearshore in such a way as to restore the initial transverse profile after the sea level rise comes to an end.

backed by high ground or mangrove swamps, but they will persist where they retreat through beach ridge plains or coastal barriers. Beaches that are already in retreat will be cut back more rapidly, but in some areas the sand supply may be maintained, or even increased, as the result of accelerated cliff erosion, or greater sediment yield from rivers because of heavier or more effective rainfall, catchment devegetation, or disturbance by tectonic uplift or volcanic activity. If the nearshore sea floor is built up by accretion, or if a continuing longshore sediment supply sustains accretion, beaches may be maintained, or even prograded, as sea level continues to rise.

Where submergence is already in progress, as on the Atlantic seaboard of the United States, a relative sea level rise has contributed to beach erosion. Bruun proposed a model of the response of a sandy beach to sea level rise in a situation where the beach was initially in equilibrium, neither gaining nor losing sand.[26] Erosion of the upper beach would then occur, with removal of sand to the nearshore zone in such a way as to restore the previous transverse profile (Figure 3-4). In effect, there would be an upward and landward migration of the transverse profile, so that the coastline would recede well beyond the limits of submergence. This restoration would be completed when the sea became stable at a higher level (see Figure 3-2), coastline recession coming to an end after a new equilibrium was achieved. The extent of recession was predicted using a formula that translates to a rule of thumb whereby the coastline retreats 50 to 100 times the amount of a rise in sea level, i.e., a 1 m rise in sea level would cause the beach to retreat by 50 to 100 m.

This model (known as the Bruun Rule) has been supported by laboratory experiments, and by changes that have occurred, for example, around the shores of the Great Lakes during episodes of rising water level.[27] If there are no gains or losses from the hinterland or alongshore it is possible to estimate how much retreat will occur, but S.C.O.R. Working Group 89 found problems in using the Bruun Rule to forecast the extent of coastline recession in response to a predicted sea level rise.[28] It is difficult to determine seaward boundary conditions for application of the Bruun Rule (i.e., the extent offshore of the profile that is to be restored at a higher level: D in Figure 3-4). Bruun[29] suggested that the boundary should be the limiting depth between predominant nearshore and offshore material, but this requires that detailed sedimentological surveys of the sea floor be available before the sea rise begins. In practice nearshore topography is variable, especially where there are sand bars, rocky outcrops, or a muddy substrate immediately offshore.

Fisher found that between 1939 and 1975 the barrier beach coast of Rhode Island retreated at an average rate of 0.2 mm/year in a period when sea level rise averaged 0.3 cm/year.[30] He calculated that 35% of the beach recession was due to losses of sand into tidal inlets and 26% by losses over the barrier islands as dunes and washover fans. This left only 39% of the beach retreat accountable through submergence and transference of beach sand seaward, as predicted by the Bruun Rule. In fact the movement of sand over and between barrier islands indicates that they are continuing the long-term landward migration, accompanied by a landward movement of lagoons and marshes, that has characterized much of the Gulf and Atlantic coastline during the Holocene.[24] Dean and Maurmeyer[31] proposed a model for the response of an entire barrier island to a phase of rising sea level, incorporating aggradation and overwash as well as adjustment of the seaward shore. As has been noted, parts of the Caspian coast are showing a similar response to sea level rise.

Beach erosion is already widespread, and only a small proportion of the world's sandy beaches are in equilibrium. The Bruun Rule cannot be applied where the beaches were initially in disequilibrium, and must also be qualified to deal with any gains or losses that occur as the result of longshore drifting and landward movement of sand from beaches into tidal inlets or onto, and across, barriers into backing lagoons and swamps. It is also necessary to allow for the time required for beaches to adjust to the changed conditions of a rising sea level, which can amount to months or years after the attainment of a new stillstand.[28] Prediction of the extent of beach recession is therefore difficult, unless a phase of sea level rise is followed by a stillstand. As long as the sea is rising, sandy coastlines will continue to recede as erosion accompanies submergence, and if the sea rises at an increasing rate, beach erosion will correspondingly accelerate. It is not possible to predict the location of a sandy coastline a century hence if sea level has risen 1 m and is continuing to rise: all that can be said with this scenario is that in a hundred years time most sandy coastlines will have retreated substantially, and that they will be eroding rapidly.

ESTUARIES AND COASTAL LAGOONS

During the worldwide late Quaternary marine transgression valley mouths were inundated to form inlets (estuaries) into which rivers flowed, and coastal lowlands were submerged to form broader embayments. The inlets at river mouths have been modified during the ensuing 6000 years of relative stillstand by sedimentation from the rivers and from the sea, which has built up bordering alluvial land and produced shoals with intervening channels. In some cases the inlets have been filled-in completely to form alluvial valley floors, some of which protrude seaward as deltas.

Alternatively, spits and barriers of sand or shingle have been built up across the mouths of inlets and embayments in such a way as to enclose coastal lagoons. These are generally shallow, and linked to the open sea by one or more tidal entrances; they are typically estuarine, with a salinity gradient increasing from the mouths of inflowing rivers to sea water at tidal entrances. This is true, for example, of the Gippsland Lakes in southeastern Australia.[23]

Coastal lagoons have generally been reduced in depth and area by accumulation of sediment washed in from rivers and from the sea, organic deposits such as peat and shells, and precipitated material sedimentation from inflowing rivers as well as by swamp encroachment during the past 6000 years.

Estuaries will tend to widen and deepen as sea level rises, with modifications of the present tidal regime and patterns of shoal deposition. The discharge of river floodwaters will be impeded as sea level rises, and a higher proportion of fluvial sediment may be retained within the submerging estuaries, instead of being delivered to the sea floor or to adjacent coastlines. There will be increasing penetration by salt water, which may also invade underground aquifers in coastal regions.

Coastal lagoons are dynamic systems that will generally be enlarged and deepened as sea level rises, with submergence and erosion of fringing shores and swamp areas, and widening and deepening of tidal entrances. Accompanying erosion of the enclosing barriers may lead to the breaching of additional entrances, and continuing submergence may eventually reopen some lagoons as marine inlets and embayments. Seawater incursion will form new lagoons in low-lying areas on coastal plains, and other such depressions will be flooded as the water table rises to form seasonal or permanent lakes and swamps.

The deepening and enlargement of coastal lagoons will be less evident where there is an abundant supply of sediment arriving at a sufficient rate to offset the effects of submergence. In southern Java the Segara Anakan is a coastal lagoon bordered by mangrove swamps that are spreading forward rapidly onto tidal mudflats, which are accreting as the result of sediment inflow from the Citanduy River. In recent decades the sediment inflow has been augmented by soil erosion, due to deforestation of the steep headwater regions. A sea level rise here will help to perpetuate this lagoon.[17]

Coastal lagoons are typically estuarine, and are likely to become more saline because of greater penetration of seawater, with an increase in their tide ranges and tidal ventilation. They will show accompanying ecological changes. Areas that were previously fresh will become brackish, with die-back of freshwater vegetation and its replacement by halophytes. Such changes have already occurred where the increase in lagoon salinity is the outcome of diminished runoff from the hinterland, as where river outflow has been intercepted by dam construction. Salinity may increase because of greater salt accessions from the subsoil, caused by vegetation clearance, augmented rainfall, or an artificially raised water table (e.g., by irrigation of adjacent farmland). Alternatively, lagoon salinity may increase after the opening or enlarging of a marine entrance, as in the Gippsland Lakes, Australia,[23] which have been modified as the result of changes in vegetation and sedimentation as brackishness increased. In such lagoons, a sea level rise will accentuate ecological changes in the already modified systems. However, a salinity increase may be offset by an increase in rainfall and freshwater runoff into the lagoon as the result of climatic change, or if additional fresh water is received as the result of river diversion.

In arid regions lagoon salinity is usually already higher than that of the open sea, and some areas have dried out as evaporite plains, as in the Coorong in South Australia. In such systems a sea level rise, producing a wider and deeper marine entrance and increased marine flushing, may actually reduce salinity.

DELTAIC COASTS

Deltas have formed where sediment brought down from the hinterland by rivers has been deposited to form low-lying land protruding into the sea. They have been built above high-tide level by sedimentation from recurrent river floods, and wave deposition along their seaward margins. A delta continues to grow where the sediment accumulating at and around the river mouth exceeds that carried away by waves and currents.

Most deltas have a history of changing positions of river mouths, with alluvial lobes developing around each mouth and then being removed by erosion after the mouth shifts. On large deltas, such as that of the Mississippi, regional isostatic subsidence under the weight of accumulating sediment has resulted in submergence of areas no longer maintained by sedimentation: successive lobes built at earlier stages are now partly or wholly submerged, and being dissected by marine erosion. Coastal subsidence has curtailed the seaward growth of most large deltas, particularly those of the Rhine, Guadalquivir, Colorado, and Amazon.

Nevertheless, many deltas are still prograding, especially in the humid tropics, where large rivers have delivered vast amounts of silt and clay to prograde the coast. In northern Java fluvial deposition around river mouths during episodes of flooding has been advancing the coastline locally by up to 200 m/year, progradation having accelerated as the result of increasing sedimentation following deforestation and the introduction of agriculture in the steep hinterland. Such deposition will accelerate if rainfall and runoff from the river catchments is augmented as a consequence of global warming. It has been estimated that the sediment yield from Javanese rivers will increase by up to 43% as the result of climatic changes expected over the coming century.[32] However, a rising sea will tend to curb the growth of deltas, and where the rate of submergence (i.e., sea level rise plus land subsidence) is greater than the rate of deposition, the coastlines will be cut back.

There has been erosion on deltaic coasts where the fluvial sediment yield has been reduced as the result of the building of dams and the impounding of sediment in reservoirs upstream. The best known example is the Nile. Here a succession of barrages built since 1902 culminated in the completion of the Aswan High Dam in 1964. The subsequent interception of Nile sediment in the Lake Nasser reservoir, behind this dam, has been followed by rapid erosion, attaining 40 m/year locally near the mouths of the Rosetta and Damietta distributaries.

As long as a delta is prograding the nearshore sea floor shows a convex profile, but when progradation comes to an end this is gradually reshaped by wave action into a concave slope that spreads landwards until it intersects the deltaic coastline. When this happens, coastal erosion suddenly becomes more rapid. This may be why erosion of abandoned delta lobes, such as that of the Hwang-Ho in China after the river mouth switched northward into the Yellow Sea in 1855, was at first gradual, and then accelerated. A rising sea, by deepening nearshore waters, will have similar effects.

SALT MARSHES AND MANGROVES

Salt marsh and mangrove vegetation has played a part in the shaping of intertidal landforms on the shores of inlets and embayments, estuaries, and lagoons. They are absent from exposed coasts, but occupy shores where wave energy is low to moderate, especially where halophytic vegetation has been able to colonize and spread across the upper intertidal zone. The growth of these plant communities has been facilitated where a large tide range produces a broad intertidal zone, and where there has been an abundant supply of sediment, especially silt and clay, together with peaty residues derived from the plants, and other organisms such as shells.

In temperate regions salt marsh vegetation has trapped sediment in such a way as to build up depositional terraces that are still growing where sediment is being supplied at a sufficient rate for their continued upward and seaward development. Salt marshes often consist of several species arranged in community zones parallel to the coastline, related to the depth and duration of regular tidal submergence. Landward they give place to freshwater vegetation such as reeds and rushes, swamp scrub, and forest, in a zonation resulting from plant succession as accreting sediment builds the substrate up to, and above, high-tide level. A rising hinterland is marked by the transition to terrigenous vegetation.

This pattern is the outcome of vertical accretion during the late Quaternary marine transgression and seaward expansion, especially after that transgression came to an end. There is often stratigraphic evidence of changes in vegetation related to this history of submergence followed by stillstand, and the seaward advance of successional zones. On coasts that are still subsiding, as in New England, salt marshes exist where they are aggrading to keep pace with the relative rise of sea level.[33] Where the aggradation is slower than sea level rise salt marshes are being drowned and eroded, and where it is more rapid than the sea level rise they are developing and spreading. Near Philadelphia, Orson[34] found that salt marshes had accreted at the rate of 1.0 to 1.2 cm/year during the past 50 years on a site where sea level had risen only 0.265 cm/year over that period.

In general, a global sea level rise will impede salt marsh progradation, and result in erosion along the seaward margins of the marshes.[35] Tidal creeks that intersect the salt marsh will be enlarged as the marsh surface is submerged. As marine submergence proceeds, the landward margins of the salt marsh can be expected to transgress onto the hinterland at a rate related to the transverse gradient; salt marsh vegetation will move landward to displace freshwater or terrigenous vegetation communities on low-lying hinterlands. Around the submerging shores of Chesapeake Bay salt marsh plants are invading the meadowland, and there are sites where the forest is dying as salt marshes encroach in Maryland[36] and behind the sounds of North Carolina.[37]

Landward migration of salt marsh communities will be impeded where the hinterland rises steeply, so that the upper intertidal zone is narrowed, and perhaps extinguished, as sea level rises. The same effect will obtain where embankments have been built at the inner margins of a salt marsh in order to reclaim land for agricultural and other uses.

Where sedimentation is sufficient to maintain vertical accretion the existing salt marshes will persist as sea level rises,[38] with landward migration, where possible, of the inner margin of the salt marsh. Reed[39] concluded that estuarine and deltaic salt marshes that continue to receive an abundant fluvial sediment supply will aggrade as sea level rises, and may increase in area, whereas salt marshes in areas of sediment deficit will become submerged and eroded.

On Baltic Sea coasts, where salinity is low, reeds and rushes form communities on sheltered sectors of the shore, spreading into shallow nearshore waters. They trap sediment and incorporate peaty material in swampland that progrades into the sea, and builds up to a level where terrigenous scrub and forest vegetation colonizes. In the northern Baltic this reed and rush encroachment has been aided by coastal emergence due to isostatic land uplift. A rising sea level will slow down this encroachment, and eventually reverse it, leading to erosion and submergence of these coastal wetlands.

Mangroves grow in tropical and subtropical environments, extending locally into the temperate zone: in the southern hemisphere they grow as far south as Corner Inlet, in Australia, at latitude 38°55′ S. In the humid tropics, mangroves grow luxuriantly, taking the place of salt marshes and forming forests that occupy the upper intertidal zone, backed by tropical rain forest or land cleared for tropical agriculture. In southeastern Asia there are more than 30 species, often zoned in distinct ecological communities, forming forests up to 40 m high intersected by branching tidal channels. On subtropical and temperate coasts, mangroves generally form a lower, scrubby seaward fringe, typically backed by salt marshes, as in Westernport Bay, Australia, with freshwater vegetation and swamp scrub to the rear. In arid or semi-arid zones the mangrove fringe may be backed by areas of unvegetated salt pan and savanna or desert hinterlands.

Like salt marshes, mangroves have become extensive during the sea level stillstand of the past 6000 years. Before that, during the late Quaternary marine transgression, they were confined to sheltered inlets and estuarine sites where they could migrate landwards, or persist on vertically accreting muddy substrates as submergence proceeded. Stratigraphic studies in northern Australia have shown that mangroves were growing in such sites as the late Quaternary marine transgression proceeded, and when it came to an end they spread out to other embayments and more exposed sites of muddy accretion along the outer coast (Figure 3-5).[40]

The effects of a sea level rise will be to reverse this sequence. In most places submergence is likely to cause die-back and erosion of the seaward margins of the

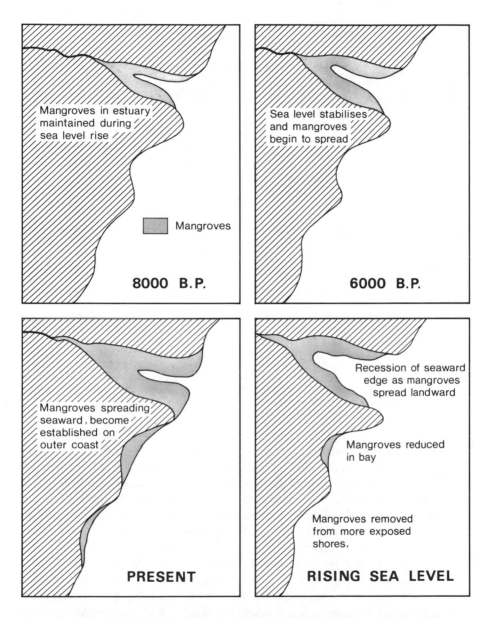

Figure 3–5 Changes in the extent of mangroves on a coastline in Holocene times. 8000 years ago, when sea level was rising, mangroves were generally confined to estuaries where mud accretion kept pace with the rising sea. As the marine transgression ended about 6000 years ago the mangroves began to spread, and now occupy embayments on the outer coastline. If sea level rises, mangroves will be reduced on the outer coastline, but may persist in muddy estuaries where accretion maintains their substrate.

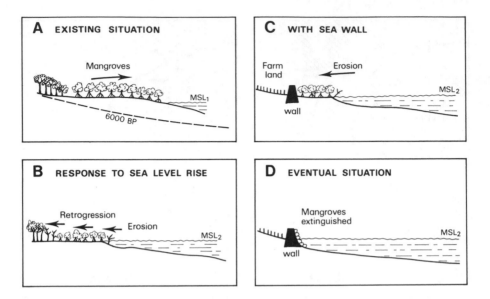

Figure 3–6 Changes on mangrove-fringed coasts as sea level rises. (A) The existing situation, mangroves typically forming a depositional terrace built by sedimentation during the past few thousand years. (B) Response to a sea level rise where nearshore water deepens and stronger wave action erodes the seaward margin of the mangroves; at the same time the rise in sea level causes retrogression of mangroves landward to displace other vegetation. (C) If a sea wall has been built at the back of the mangroves to reclaim and protect farmland or other developed land to the rear the mangrove fringe will be narrowed as sea level rises. (D) Eventually the sea wall may have to be armoured against increased exposure to wave attack.

mangroves, and they will tend to migrate landward as the sea rises if low-lying hinterland sites are available (Figure 3-5). Mangroves will disappear from the more exposed areas, and retreat behind the more sheltered inlets and embayments: as the seaward margin is cut back, the landward margin will migrate to displace backing vegetation (Figure 3-6 A, B). Under natural conditions mangroves are backed by low-lying estuarine and alluvial land, with freshwater swamp or forest, which they will invade as a rising sea drives the mangrove zone landward, but in many areas the inner margin is defined by a wall built to protect reclaimed land.

Evidence of changes in mangroves has been documented on coasts where the sea level is rising because of land subsidence. At Port Adelaide in South Australia, where local tide gauge records indicate that a relative rise of sea level has been taking place, mangroves spread back into salt marshes at the rate of 17 m/year between 1935 and 1979.[41] Similar changes have been mapped on the marsh islands of Corner Inlet in southeastern Australia.[42]

As with salt marshes, mangroves will persist only where the rate of substrate accretion equals or exceeds the rate of sea level rise, particularly near the mouths of rivers.[43] There is also a possibility that as the sea rises, sediment stirred from bordering mudflats by wave action will be washed up into the mangrove fringe, raising the substrate. Such a process could maintain, or even widen, the mangrove fringe as its inner margin transgressed landwards, but the extent to which this will occur is a topic for further research. In some places, such as the West Johore coast in Malaysia, the seaward margin of the mangroves has been cut back by erosion, so that they persist on a terrace with a sharp cliff along the seaward side. Onshore movement of muddy sediment is impeded where such a cliff has formed because this cliff reflects wave action and promotes scouring of the fronting mudflats.

Mangroves are unlikely to survive on low (coral) islands as sea level rise proceeds because of the low rates of sediment accretion in areas remote from rivers and terrigenous sources.[44]

Mangroves have been extensively modified by the impact of human activities, especially during the last few decades. Some have been exploited for timber production and fuel wood; others have been cleared in order to dredge out placer deposits, such as tin, which occur beneath the mangrove mud, but the most extensive clearance has been made for the establishment of aquaculture (fish and prawn ponds), and salt pans. In the vicinity of urban centers, mangroves have been reclaimed for urban expansion and industrial and port development. Where the mangrove area has been converted to aquaculture, a sea level rise will threaten to breach the enclosing banks and submerge the fish or prawn ponds. If these ponds are to be maintained the enclosing walls and the floors will have to be raised to match the levels of the rising sea. Once a protective sea wall has been built, a rising sea level is likely to narrow and extinguish the mangrove fringe that lies seaward (Figure 3-6 C, D).

Where mangroves have been reclaimed for agriculture, usually rice farming or palm plantations producing rubber, oil, or coconuts, embankments (bunds) have been built at the inner margin of the remaining mangroves. Where the rear of the mangroves is thus delimited, attempts will probably be made to maintain it as sea level rises, raising it to prevent wave overtopping and marine flooding. If this happens, the retreating mangrove fringe will not be able to colonize the hinterland; it will be reduced, and in many places will disappear altogether as the intertidal zone narrows and steepens. The coastline will thus become more and more artificial.

INTERTIDAL FLATS AND NEARSHORE AREAS

Tidal mudflats and sandflats generally lie seaward of salt marshes or mangroves, and are unvegetated except for a relatively sparse and variable cover of seagrasses and marine algae. They slope gradually down to low-tide level, and are usually traversed by tidal creeks and seepage corridors of soft, wetter sediment. In some areas they contain shelly organisms, such as cockles, mussels, and oysters, and shelly debris derived from these can form gravelly shoals and backing beaches.

As sea level rises the outer part of the intertidal flats will become permanently submerged, but erosion of salt marshes, mangrove swamps and the coastal lowland fringe will extend it landward. Where sea walls have been built, however, tidal mudflats, like salt marshes and mangroves, will be reduced and eventually obliterated by the rising sea. Substantial areas of tidal mudflats have already been enclosed by sea walls for land reclamation in southeastern Asia, and there are plans to reclaim many more areas. Such reclamation will further increase the extent of artificial coastlines.

Seagrasses and associated plants, notably marine algae, occupy parts of the intertidal and nearshore zones. Seagrasses require a suitable substrate, generally mud or fine sand, and light penetration generally limits them to low-tide depths of less than 10 m. Their growth is improved by accessions of fine-grained sediment and the increase in nutrients that results from nontoxic pollutants. However, if the nutrients become excessive (eutrophication) seagrass beds can be blanketed and destroyed by proliferating algae. Seagrasses promote organic accretion, and can diminish wave action, thereby helping to stabilize the adjacent coast. Where the seagrasses have been destroyed wave action is less impeded, and waves attack the coastline more strongly. Marine algae have generally similar effects, but can grow on coarser sediment and rocky outcrops, and extend into water tens of meters deep.

As sea level rises the intertidal and nearshore zones occupied by seagrasses and marine algae will tend to migrate landwards, the inner margin spreading on to the submerging sandy and muddy substrates formed as beaches are submerged and salt marshes and mangrove swamps overwashed, the outer margin dying away as the water becomes too deep. It is possible that the seagrass zone will actually become wider and richer on coasts where broad plains, rich in soil nutrients, are being submerged by the rising sea.

CORAL REEFS

Corals and associated organisms, notably calcareous algae, have built extensive reefs in tropical seas, especially in the western parts of the oceans within a zone extending 30° N and 30° S of the Equator. They include fringing reefs built out from the coastline, barrier reefs built offshore, parallel to the coastline and separated from it by a lagoon, atolls encircling a marine lagoon, and outlying patch reefs. Each has been built by coral polyps, small marine organisms that extract carbonate from seawater and grow into a skeletal framework, which is infilled by algae, shells, and other organisms as well as detrital and precipitated carbonates, to form a reef structure.

Corals grow best in clear, warm seawater with temperature exceeding 18°C and salinity in the range 27 to 38 parts per thousand. Their growth is impeded when the temperature exceeds 28°C, where the sea becomes hypersaline, as in parts of Shark Bay, western Australia, or where it is diluted by fresh water off river mouths.

Coral reefs have grown up from depths of at least 50 m to form platforms exposed at low tide where, as corals cannot survive prolonged exposure to the atmosphere, they die. In places reefs have been built up to just above low-tide level by organic growth and associated sedimentation, the surfaces exposed at low-tide being dominated by dead corals and living algae. Coral growth remains vigorous on the steep flanks of reefs below low tide level, especially on windward shores, washed by strong well-aerated surf, where the nutrient supply is relatively abundant. The most extensive areas of upward growing coral are on the lee side of barrier reefs, in atoll lagoons, and in waters previously shallowed by sedimentation.

Some reef flats are surmounted by small islands (cays) of coralline sand and gravel eroded from the surrounding reef and washed up by wave action. They rise only a meter or two above high-tide level, and carry land vegetation, usually modified by human settlers.

Many coral reefs have already been damaged by the impacts of human activities. These have included the use of explosives to harvest fish, and the beating of shallow reef areas to drive fish into nets. The use of chemicals and gases to capture fish has also had adverse ecological consequences. Some reefs have been quarried for sand, gravel, and building stone, or been damaged by the cutting of boat access channels, or by boat anchors. Many have been affected by pollution, including the in-washing of muddy sediment generated by nearby dredging or drilling for oil. In recent decades, deforestation of hinterlands has increased turbidity in coastal waters, killing many fringing and nearshore coral reefs by blanketing them with sediment. Excessive nutrients from eroding soils, agricultural fertilizers, and sewage pollution also cause eutrophication detrimental to coral growth. Collecting of shells and precious corals has impoverished many reefs, and outbreaks of destructive crown-of-thorns starfish on coral reefs may be a response to human activities, although some have interpreted this as a natural phenomenon.

The effects of a rising sea level on coral reefs depend on the ability of corals to grow on the submerging reef flats.[45] Living corals are few and scattered on reefs, and if they are killed they are not quickly replaced. As has been indicated, many corals are already under various kinds of ecological stress as the result of human activities, and some of the less vigorous may fail to revive, so that the reefs become permanently submerged as sea level rises. Fringing and nearshore reefs are less likely to survive than outlying reefs because of increasing turbidity in coastal waters, especially where runoff from the hinterland increases as the result of climatic change, and as the result of larger waves eroding the coast.

Where a slowly rising sea level stimulates the revival of coral growth on reef flats it is possible that coral communities will maintain their level relative to the rising sea.[46] That corals can respond in this way is shown by the local revival of coral growth as the result of "moating" of water levels behind shingle ramparts or algal rims, where a slightly higher veneer of living coral is found in the form of micro-atolls, without any sea level change.[47] Cubit[48] found that the alga *Laurencia papillosa* migrated across coral reefs at Punta Galena, in Panama, in response to annual fluctuations of local mean sea level. These local occurrences emphasize the lack of any general revival of corals in response to the supposed contemporary sea level rise.[49]

There is great variation in rates of coral growth, but a review by Buddemeier and Smith indicated that under favorable conditions corals could grow upward 10 mm/year.[50] Measurements of coral growth are generally in the range 0.4 to 7 mm/year,[51] and studies of tilting coral reefs, such as Uvea in the Loyalty Islands, have shown zones of slow submergence (<5 mm/year) where upward growth is being maintained passing laterally to zones of more rapid submergence (5 to 10 mm/year) where the corals are growing but failing to keep up with the rising sea, and zones (>10 mm/year) where the reefs are drowned and inert (Figure 3-7).

CHANGES ON A TILTED CORAL REEF

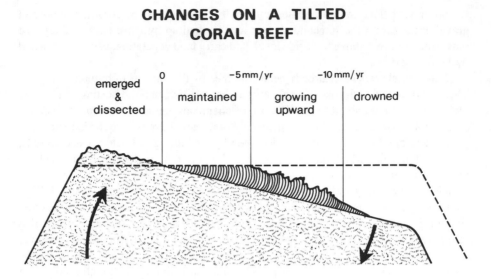

Figure 3–7 The response of coral reefs to a sea level rise may be predicted with reference to changes on tilting reefs or atolls. In this example the reefs are being maintained by upward growth of corals, associated organisms and sedimentation in the zone subsiding at less than 5 mm/year, are still growing upward in deepening water in the zone subsiding at 5 to 10 mm/year, but are drowned where the subsidence rate exceeds 10 mm/year.

Where low islands (cays) have formed on reef flats they are likely to be eroded, and may disappear, overwashed by storm surges, as sea level rises. This is a matter of great concern in populated oceanic islands such as the Maldives.[52] They could survive where submergence is slow enough for the corals to grow upward and maintain a surrounding protective reef, and in these conditions could even be enlarged by accretion of coralline material derived from the growing reef.[49]

ARTIFICIAL COASTS

There are many examples around the world of damage and destruction of coastal property, including buildings, roads, farmed land, and forestry areas, as the result of cliff recession or the erosion of beaches, deltas, salt marshes, and mangrove swamps. Large sums of money have been spent by governments and local authorities on the building and enlarging of concrete sea walls and boulder ramparts where erosion threatened developed land in coastal urban areas, especially where beach resorts and tourist facilities have been developed on low-lying sandy coasts.[53] Artificial coasts have also developed where land reclamation has required the building of an enclosing sea wall. In southeastern Asia extensive areas of low-lying coast have been modified for the development of fish and prawn ponds, and are now enclosed from the sea by low artificial embankments.

Sea walls and boulder ramparts usually cause wave reflection, which depletes adjacent beaches. This is particularly unwelcome at seaside resorts where the beaches were the original tourist attraction. An alternative method of shore protection is artificial beach nourishment, which has the advantage of providing or restoring a coastal recreational resource. Unfortunately it is expensive, and may only be feasible in a few intensively developed urban resort areas, but in suitable areas it will be possible to maintain beaches as sea level rises.[21]

Existing structures have been built in relation to present sea level. As sea level rises sea walls and boulder ramparts are likely to be raised and extended laterally, and new structures will be built to combat erosion and prevent the ingress of the sea on to developed agricultural, industrial, or urbanized land. This has happened repeatedly along the subsiding coastline of the Netherlands, where more than half the country is now below the level of high tides in the North Sea. Were it not for successive enlargement of protective sea walls, much of the Netherlands would now be a large estuarine wetland. A global sea level rise is likely to prompt the building of many more sea walls, and the prospect is that low-lying sectors of the world's coastline will become as artificial as parts of the Netherlands.

CONCLUSIONS

Coastal landforms have evolved in relation to present sea level, which has been relatively stable around much of the world's coastline over the past 6000 years. They continue to change in response to geomorphological processes and ecological conditions. On subsiding coasts the changes have already been complicated by the effects of a relative sea level rise, and these coasts yield evidence of what will happen during a global sea level rise, expected as a consequence of the enhanced greenhouse effect. Attempts are being made to develop theoretical models to predict the consequences of a sea level rise, but these are impeded by inadequate data on existing rates of change, process systems, and sediment budgets. The scenario of an accelerating sea level rise poses difficult problems because of the delays in response completion of landforms to processes. Predictions of the extent of coastline change can really only be made when there is the prospect of attainment of a new sea level stillstand.

REFERENCES

1. Kidson, C. Sea level changes in the Holocene, In O. van de Plassche, Ed. *Sea-Level Research: A Manual for the Collection and Evaluation of Data* (Geo Books, Norwich, 1986), pp. 27–64.
2. Komar, P. D. The 1982–83 El Niño and erosion on the coast of Oregon, *Shore and Beach*, 64: 3–12 (1986).

3. Gutenberg, B. Changes in sea level, postglacial uplift, and the mobility of the Earth's interior, *Bull. Geol. Soc. Am.* 52: 721–772 (1941); See also Barnett, T. P. The estimation of 'global' sea level changes: a problem of uniqueness, *J. Geophys. Res.,* 89: 7980–7988 (1984).

4. Gornitz, V. and S. Lebedeff. Global sea level changes during the past century, In D. Nummedal, O. H. Pilley, and J. D. Howard, Eds., *Sea-Level Change and Coastal Evolution* (Society of Economic Paleontologists and Mineralogists 1987), Special Publication 41, pp. 3–16, See also Fairbridge, R. W. Mean sea level changes, *Encyclopaedia of Oceanography*, (Reinhold, New York, 1966), pp. 479–485.

5. Pirazzoli, P. A. Secular trends of relative sea-level changes indicated by tide-gauge records, *J. Coastal Res. Special Issue,* 1: 1–26 (1986).

6. Mörner, N. A. Eustacy and geoid changes, *J. Geol.,* 84: 123–151 (1976).

7. Bird, E. C. F. and K. Koike. Man's impact on sea level changes: a review, *J. Coastal Res. Special Issue* 1, 83–88 (1986).

8. Lewis, J. R. *The Ecology of Rocky Shores.* (English Universities Press, London, 1964).

9. Wanless, H. R. Sea level is rising - so what?, *J. Sedim. Petrol.,* 52: 1051–1054 (1982).

10. Bird, E. C. F. The tubeworm *Galeolaria caespitosa* as an indicator of sea level rise, *Victorian Nat.,* 105: 98–104 (1988).

11. Barth, M. G. and J. G. Titus, Eds. *Greenhouse Effect and Sea Level Rise* (Van Nostrand Reinhold, New York, 1984).

12. Houghton, J. T., G. J. Jenkins, and J. J. Ephraums, Eds, *Scientific Assessment of Climate Change.* (Cambridge University Press, 1990).

13. Leontiev, O. K. and K. A. Veliev, *Transgressive Coasts.* International Symposium, Baku, USSR, 15–21 September 1990.

14. Bird, E. C. F. *Coastline Changes* (Wiley Interscience, Chichester, 1985).

15. Mehta, A. J. and R. M. Cushman, Eds., *Proceedings of the Workshop on Sea Level Rise and Coastal Processes* (Department of the Environment, Florida, 1989).

16. Bird, E. C. F. Physiographic indications of a sea-level rise, In G. I. Pearman, Ed, *Greenhouse* (Brill, Leiden, 1988), pp. 60–73.

17. Bird, E. C. F. and O. S. R. Ongkosongo. *Environmental Changes on the Coasts of Indonesia* (United Nations University, 1980)

18. Sunamura, T. Projection of future coastal cliff recession under sea level rise induced by the Greenhouse Effect: Nii-jima Island, Japan, *Trans. Jpn Geomorph. Union*, 9 (1): 17–33 (1988).

19. Bird, E. C. F. and N. J. Rosengren, Coastal cliff management: an example from Black Rock Point, Melbourne, Australia, *J. Shoreline Manage.,* 3: 39–51 (1987).

20. Pirazzoli, P. A. Marine notches, In O. Van der Plassche, Ed., *Sea-Level Research: A Manual for the Collection and Evaluation of Data,* (Geo Books, Norwich, 1986), pp. 361–499.

21. Schwartz, M. L. and E. C. F. Bird, Eds., *Artificial Beaches*, Journal of Coastal Research, 1990, Special Issue 6.

22. Schwartz, M. L., Ed., *Barrier Islands.* (Dowden, Hutchinson & Ross, Stroudsburg, Pennsylvania, 1973).

23. Bird, E. C. F. *A Geomorphological Study of the Gippsland Lakes* (Ministry for Conservation, Melbourne, Publication 186, 1978).

24. Leatherman, S. P., Ed,, *Overwash Processes.* (Dowden, Hutchinson, and Ross, Stroudsburg, Pennsylvania, 1981).

25. Bird, E. C. F. The world's disappearing beaches, in *Science and the Future Handbook*, (Encyclopaedia Britannica, 1987), pp. 115–127.

26. Bruun, P. Sea level rise as a cause of shore erosion, *Proc. Am. Soc. Civil Eng. (Waterways and Harbors Division)*, 88: 117–130 (1962).

27. Schwartz, M. L. The Bruun theory of sea level rise as a cause of shore erosion, *J. Geol.*, 75: 76–92 (1967).

28. S.C.O.R. Working Group 89. The response of beaches to sea-level changes: a review of predictive models, *J. Coastal Res.*, 7 (3): 895–921 (1991), See also Leatherman, S. P. Modelling shore response to sea-level rise on sedimentary coasts, *Prog. Phys. Geogr.*, 14 (4): 447–464 (1990).

29. Bruun, P. The Bruun Rule of erosion by sea level rise: a discussion of large-scale two- and three-dimensional usages, *J. Coastal Res.*, 4: 627–648 (1988).

30. Fisher, J. J. Regional long-term and localized short-term coastal environmental geomorphology inventories, In J. E. Costa and P. J. Fleischer, Eds., *Developments and Applications of Geomorphology*. (Springer-Verlag, Berlin, 1984), pp. 68–96.

31. Dean, R. G. and E. M. Maurmeyer, Models for beach profile response, In P. D. Komar, Ed., *Handbook of Coastal Processes and Erosion* (CRC Press, Boca Raton, FL, 1983), pp. 151–165.

32. Parry, M., Ed., *The Potential Impact of Climate Change in Southeast Asia*. (United Nations Environment Programme, Bangkok, 1990).

33. Redfield, A. C. Development of a New England salt marsh, *Ecol. Monogr.*, 42: 201–237 (1972).

34. Orson, R., W. Panagetou, and S. P. Leatherman. Response of tidal salt marshes of the United States Atlantic and Gulf coasts to rising sea levels, *J. Coastal Res.* 1: 29–37 (1985).

35. Phillips, J. D. Coastal submergence and marsh fringe erosion, *J. Coastal Res.*, 2: 427–436 (1986).

36. Darmody, R. G. and J. E. Foss. Soil-landscape relationships of the tidal marshlands of Maryland, *J. Soil Sci. Soc. Am.*, 43: 534–541 (1979).

37. Stevenson, J. C., L. G. Ward, and M. S. Kearney. Vertical accretion in marshes with varying rates of sea level rise. In D. A. Wolfe, Ed., *Estuarine Variability*. (Academic Press, Orlando, Florida, 1986), pp. 241–259.

38. Pethick, J. S. Long-term accretion rates on tidal salt marshes, *J. Sedim. Petrol.*, 51: 571–577 (1981).

39. Reed, D. J. The impact of sea-level rise on coastal salt marshes, *Prog. Phys. Geogr.*, 14 (4): 465–481 (1990).

40. Woodroffe, C. D., B. G. Thom, and J. Chappell. Development of widespread mangrove swamps in mid-Holocene times in northern Australia, *Nature*, 317: 711–713 (1985).

41. Burton, T. Mangrove changes recorded north of Adelaide, *Safic*, 6: 8–12 (1982).

42. Vanderzee, M. P. Changes in saltmarsh vegetation as an early indicator of sea-level rise. In G. I. Pearman, Ed., *Greenhouse* (Brill, Leiden, 1988), pp. 147–160.

43. Woodroffe, C. D. The impact of sea-level rise on mangrove shorelines, *Progr. Phys. Geogr.*, 14 (4): 483–520 (1990).

44. Ellison, J. C. and D. R. Stoddart. Mangrove ecosystem collapse during predicted sea-level rise: Holocene analogues and implications, *J. Coastal Res.*, 7 (1): 151–165 (1991).

45. Stoddart, D. R. Coral reefs and islands and predicted sea-level rise, *Prog. Phys. Geogr.*, 14 (4): 521–536 (1990).

46. Neumann, A. C. and I. Macintyre. Reef response to a sea level rise: keep-up, catch-up or give-up, *Proc. 5th Int. Coral Reef Congr.*, 3: 105–110 (1988).

47. Hopley, D. *The Geomorphology of the Great Barrier Reef.* (Wiley Interscience, New York, 1986).

48. Cubit, J. D. Possible effects of recent changes in sea level on the biota of a Caribbean reef flat, and predicted effects of rising sea levels, *Proc. 5th Int. Coral Reef Congr.,* 3, 111–118 (1985).

49. Bird, E. C. F. The impacts of sea level rise on coral reefs and reef islands, In A. Vallega, Ed., *International Conference on Ocean Management in Global Change,* Genoa, Italy, June 1992: in press.

50. Buddemeier, R. W. and S. V. Smith. Coral reef growth in an era of rapidly rising sea level: predictions and suggestions for long-term research, *Coral Reefs,* 7: 51–56 (1988).

51. Hopley, D. and D. W. Kinsey. The effects of a rapid short-term sea-level rise on the Great Barrier Reef, In G. I. Pearman, Ed., *Greenhouse* (Brill, Leiden, 1988), pp. 189–201.

52. Pernetta, J. C. and G. Sestini. *The Maldives and the Impact of Expected Climatic Changes.* UNEP Regional Seas Report and Studies, 104. (UNEP, Nairobi, 1989).

53. Walker, H. J. *Artificial Structures and Shorelines* (Kluwer Academic Publications, 1988).

CHAPTER 4

River Flux to the Sea: Impact of Human Intervention on River Systems and Adjacent Coastal Areas

John D. Milliman and Ren Mei-e

ABSTRACT: Rivers presently discharge about 35×10^3 km^3 of freshwater and 15 to 20×10^9 tons of sediment to the world oceans annually. More than half of the water and about 75% of the sediment come from Oceania and southern Asia. Most existing budgets for river sediment, however, have underestimated the sediment fluxes from small mountainous rivers, perhaps by as much as a factor of three. In contrast, sediment fluxes to the ocean from large rivers have been overestimated, as some of the sediment load is subaerially sequestered in subsiding deltas and within estuaries.

Prior to shifting land-use patterns, the quantity and distribution of sediment discharge was signficantly different than at present. Widespread farming and deforestation (beginning 2000 to 2500 years ago) increased sediment loads of rivers significantly, locally by an order of magnitude, particularly in southern Asia and Oceania, where the source rocks are erodable and seasonal precipitation is particularly heavy. In contrast, increased river diversion, particularly dam-building, has resulted in recent decreases in local sediment flux to the oceans, especially in Europe and North America.

In the near future terrestrial erosion rates probably will continue to increase (particularly in Asia), but the transfer of both water and sediment to the sea almost certainly will decrease as more dams and other river projects are built and utilized. Low-lying deltas may be particularly vulnerable in the near future, as their rapid buildup over the past several millenia appears to have been due in part to anthropogenically enhanced terrestrial erosion and river discharge, which soon will be negated by construction of dams and other river diversions. Effects from the decreased flux of fresh water, such as shrinking mangrove forests and decreased coastal fisheries, probably will be felt immediately, but coastal erosion may lag considerably. Uncompensated (or even accelerated) subsidence of low-lying deltas may have a much greater local impact than the global rise of sea level.

INTRODUCTION

Rivers represent one of mankind's most widely used and beneficial natural resources. River waters irrigate, both naturally and artificially, vast areas of the earth's surface; nutrient-rich waters form a prime source for the high biological productivity that characterizes many estuarine and coastal waters; and the river itself furnishes a natural pathway to the interior of many land areas. But rivers also are the source of disastrous floods that can devastate natural and human habitats; it is not without reason that Chinese refer to the Yellow River (Huanghe) as "source of joy, source of sorrow".

With increasing demographic pressures, however, rivers and their deltas are being stressed by rapid modification and urbanization. For example, at present more than half of the megapolises in the world (urban populations greater than 5 million) are located in southern Asia, and of these all but one (Beijing, China) are located at least partly on rivers and/or coastal lowlands (Figure 4-1). Combined with rapid economic expansion throughout much of Asia and the Pacific, this situation assures increased future environmental stress (i.e., deforestation, demise of the mangrove forests, river diversions, and beach erosion) that will augment or surpass the envisioned effects of rising global sea level.

As river basins and their deltas become more densely populated, the need for river regulation increases. Containment of annual floods, particularly where rainfall is periodically heavy, is one reason, but hydroelectric power and irrigation also are important. The less water that is "lost" to the ocean, the more can be utilized by land-managers and farmers. The problem, of course, is that river containment may have many consequences not experienced by untrained rivers. The decreased flux of sediment and fresh water to the ocean, for instance, has had dramatic effects on the coastal environment (see below), and it is predicted that future changes will only exacerbate these trends globally.

This chapter discusses the global picture of water and sediment discharge from rivers to the ocean. We then discuss ways in which present-day rivers affect and are affected by human activities and the ways in which rivers will change their character in response to future climatic and anthropogenic changes. We discuss several examples of large rivers that have been or are being dammed and diverted.

Much of our interest ultimately focuses on the coastal zone, since it has greater impact on human activities than any other marine environment, but the reverse is also

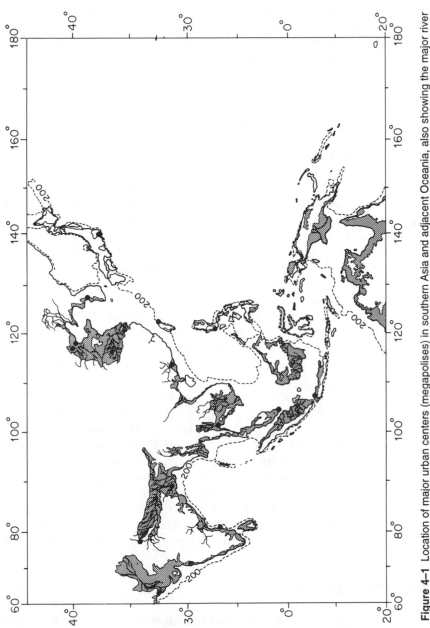

Figure 4–1 Location of major urban centers (megapolises) in southern Asia and adjacent Oceania, also showing the major river systems and low-lying coastal areas (shaded). Note that nearly all megapolises lie near or within zones that could be impacted by river diversion or coastal change resulting from that river diversion.

true. The management of the coastal zone therefore must take into account both the effect of human activities as well as impacts resulting from corresponding changes in the coastal zone. It is hoped that this discussion will give the reader a sense of the types of impacts that alteration of a river basin, both increased erosion on the one hand and damming/diversion on the other, can have on the viability of a fluvial/ estuarine system. Clearly, these problems (and their possible mitigation) should be kept in mind when discussing or planning future river management.

RIVER SYSTEMS

WATER DISCHARGE

Predicting changes in fluvial processes and possible environmental impacts of such changes assumes knowledge of the river flow and sediment load as well as their discharge to the sea. Water flow is generally a simple function of the net precipitation (i.e., precipitation minus evaporation) and basin size. To compare flow in large and small rivers, many workers refer to runoff, which is discharge normalized over the river basin area. Runoff generally equates directly to net precipitation.

Water discharge to the sea is estimated at somewhere between 33 and 39×10^3 km^3/year,[1,2] which is slightly less than the 40×10^3 km^3 net terrestrial precipitation calculated by Baumgartner and Reichel.[3] As Baumgartner and Reichel's calculations include climatic data from the Arctic and Antarctic, from which there is no river discharge, it is not surprising that their values are higher than the calculated river discharge. Also, at least some of the terrestrial precipitation enters the sea via groundwater discharge.

In terms of worldwide water discharge, the world's ten largest rivers account for about 38% of the total fluvial water entering the ocean, slightly greater than their combined percentage of drainage basin area (Table 4-1). The Amazon River alone contributes nearly 20% of the world total (about 6300 km^3/year), more than the combined total of the next seven largest rivers! Not surprisingly, tropical areas with heavy rainfall — specifically southern Asia, Oceania, and northeastern South America — are the prime contributors: about 65% of the global total (Figure 4-2). In contrast, with the exception of the Zaire and Niger Rivers, Africa contributes practically no fluvial water to the oceans. Input from the Eurasian Arctic is significant only because of the large area drained by these north-flowing rivers, net precipitation in this area being very low.

SEDIMENT DISCHARGE

The sediment transport of a river is a first-order function of river basin size and topography (in reality a surrogate for tectonism) (Figure 4-3); in most river basins other factors, such as climate, geology, and human activity, are second-order factors.[4] Although the sediment load of a river decreases with decreasing basin area, the yield

Table 4–1 Ranking of the World's Major Rivers in Terms of Drainage Area, Flow, and Sediment Load

Drainage Basin Area

River	$\times 10^6 km^2$
1. Amazon	6.15
2. Zaire	3.72
3. Mississippi	3.27
4. (Nile)	(3.03)
5. Parana	2.60
6. Yenissei	2.58
7. Ob	2.49
8. Lena	2.49
9. Changjiang	1.91
10. Amur	1.86
11. MacKenzie	1.81
	(30% of world total)

Average Annual Water Discharge

River	km^3/year
1. Amazon	6300
2. Zaire	1250
3. Orinoco	1100
4. Ganges/Brahmaputra	971
5. Changjiang	921
6. Mississippi	580
7. Yenissei	560
8. Lena	514
9. Plata	470
10. Mekong	470
	(33% of world total)

Average Annual Suspended Load

River	Mt
1. Amazon	1200
2. Huanghe	1080
3. Ganges/Brahmaputra	1050
4. Changjiang	480
5. Irrawaddy	260
6. Magdalena	220
7. Mississippi	210
8. Godvari	170
9. Mekong	160
10. Orinoco	150
	(30% of world total)

Data from Milliman and Meade,[7] modified slightly with new compilations from Milliman and Syvitski.[4] The Nile drainage basin is listed even though it presently is nearly completely dammed.

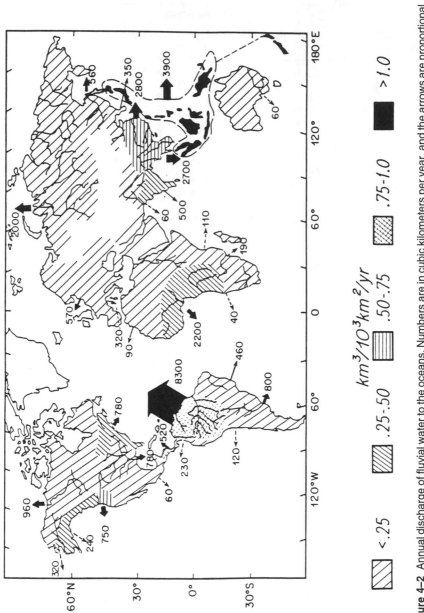

Figure 4–2 Annual discharge of fluvial water to the oceans. Numbers are in cubic kilometers per year, and the arrows are proportional to the numbers. (From Milliman-J. D., 1991. In R. F. C. Mantoura, J. M. Martin, and R. Wollast, Eds., *Ocean Margin Processes in Global Change*. John Wiley and Sons Ltd., pp. 69–89. With permission.)

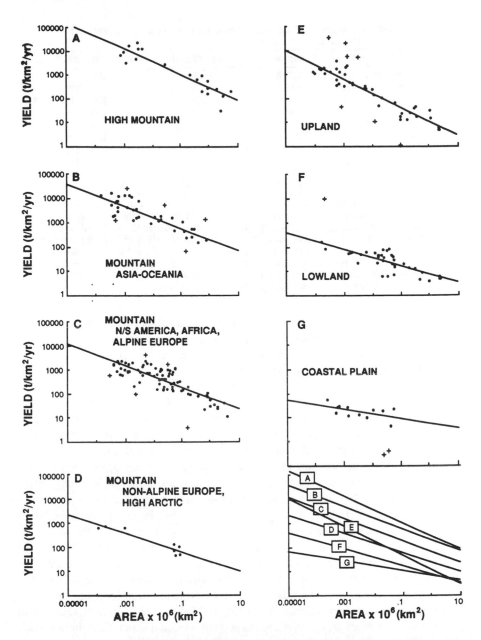

Figure 4–3 Variation of sediment yield (sediment load normalized for drainage basin area) for seven topographic categories of river basins. For all river types except lowland and coastal plain rivers, the correlation is strong, r^2 ranging from 0.70 to 0.89. (From Milliman, J. D. and Syvitski. J. P. M., 1992. *J. Geol.*, 100, 525–544. With permission.)

Table 4–2 Comparison of Calculated Suspended Loads from Various Land Areas

Area (×10^{6}km^2)		Holeman[5]	Milliman/Meade[7]	Milliman/Syvitski[4]	
N/C America	(18)	1,780	1,462	>1,500	
S. America	(18)	1,090	1,788	3,000	(?)
Europe	(5)	290	230	500	(?)
Africa	(15)	490	530	800	(?)
Eurasian Arctic	(11)	—	84	84	
Asia	(17)	14,480	6,349	<6,400	(?)
Oceania	(3)	—	3,000	9,000	(?)
Australia	(2)	210	62	100	(?)
Total		18,340	13,505	~ 20,000	

Note that while the totals between Holeman[5] and Milliman and Syvitski[6] are similar, the fluxes from various land areas differ greatly. Holeman assumed that the sediment flux from Asia could be estimated by extrapolating measured loads of various Asian rivers over all of Asia, even though much of Asia (i.e., the Eurasian Arctic) is characterized by rivers with very low sediment yields. Milliman and Meade[7] first recognized the importance of rivers draining the high-standing islands of Oceania, but underestimated their actual input to the ocean. Ironically, Holeman[5] ignored the sediment input from these islands.

usually increases because smaller basins have less flood plain area in which sediment can be stored (i.e., deposited). Thus, a large river like the Amazon may only discharge a small percentage of its sediment to the ocean (its sediment yield at Obidos is 190 t/km^2/year); in contrast, the small Peinan (Taiwan) has only $^1/_{400}$ the basin size of the Amazon, but it has a sediment yield of 14,800 t/km^2/year! Small river basins also are affected by catastrophic events, which can increase river sediment loads by one to several orders of magnitude. In contrast, most large river basins tend to modulate peak events such that a major flood in one part of the basin often has relatively little effect on the sediment discharge of the entire river.

Athough the exact amount of sediment discharged to the global ocean from present-day rivers is not known precisely, most estimates fall within the range of 15 to 20 × 10^9 t/year.[4,5,6,7] With the exception of the Chinese rivers, most large rivers in Asia and South America are poorly documented, whereas many rivers in North America and Europe are well studied but have small sediment loads. The similarity in the estimates by Holeman[5] and Milliman and Syvitski,[4] however, belies the marked regional differences in their numbers (Table 4-2). Holeman[5] concluded that Asia (including Eurasia) accounted for substantially more than half the sediment to the sea, while Oceania (Indonesia, Philippines, Taiwan, New Guinea, and New Zealand) was not even considered. According to Milliman and Syvitski,[4] about half of the sediment carried by rivers is discharged from Oceania, and if southern Asia is included, this estimate increases to about 75%. The dominance of rivers draining Oceania is a result of their generally small basin size and the cumulative effect of heavy rainfall, erodable soils, and human impact from deforestation and agriculture.

If the input of sediment from small rivers has been underemphasized by most workers, the discharge from large rivers probably has been overemphasized. Most large rivers are gauged sufficiently upstream from the river mouth to eliminate tidal effects, and much of this sediment is deposited on flood plains and on subaerial parts of the deltas, particularly during annual floods. Re-evaluation of large rivers now indicates that perhaps as little as half of the sediment gauged in some of the large rivers actually reaches the ocean (Meade and Milliman, unpublished data). Presumably the more extensive the subsiding delta over which a river flows, the less sediment reaches the ocean; for example, 40 to 60% of the sediment gauged upstream on the Brahmaputra and Ganges rivers may be deposited over the subsiding delta and not reach the Bay of Bengal (Milliman and Bruk, unpublished data). Because most large rivers (in terms of sediment discharge) are located on nonactive continental margins, the adjacent shelf is relatively wide. As a result, little sediment appears to escape the inner shelf during high-stands of sea level. Considering the Ganges-Brahmaputra, Huanghe, and Amazon, the three largest rivers in terms of sediment loads, for instance, perhaps no more than 10 to 20% of their combined loads escapes the nearshore and inner shelf (Meade and Milliman, unpublished data).

The significance of these statements concerning the fate of fluvial sediment can be seen when discussing the impact of dams on coastal systems. If the sediment carried by a river can escape the coastal zone (i.e., bypasses to the deep sea, as active margin rivers often do), then trapping of sediment in reservoirs behind dams may have less negative impact than trapping the sediment from a large river that normally remains in the coastal zone. It is in this latter case that such problems as coastal erosion and uncompensated subsidence can become critical in coastal zone management. These points are discussed at greater length in following sections.

ATHROPOGENIC CHANGES IN RIVER SYSTEMS AND THEIR ENVIRONMENTAL EFFECTS

Human activities can affect the discharge of water and sediment from a river in many ways. Deforestation and agricuture, as well as urbanization, can increase the erosion of the river basin by as much as an order of magnitude.[8] In tropical areas alone, 100,000 km^2 are being deforested annually.[9] Freshly exposed soil is much less likely to retard erosion by rainfall or moving water, particulary in those areas where cleared land is used for agriculture and the rainfall is heavy.[10] The very high sediment loads in modern Asian rivers reflect a considerable influence from human activities, particularly poor agricultural practices in conserving soil. The Yellow River, for example, transports an order of magnitude more sediment than it did prior to widespread cultivation of the loess plateaus in northern China about 2400 years ago.[11] One implication of this large increase in the sediment discharge of Asian rivers has been the increased shoreward accretion of deltaic areas over the past several millenia. Shanghai, with an urban population approaching 20 million, for example, was intertidal as recently as 2000 years ago. Thus, mankind in Asia is occupying land areas that may not have existed if not for increased upstream erosion.

In contrast, diverting or damming a river can interrupt or completely stop the flow of water and sediment to the sea. Several examples of damming are well known — e.g., the complete cessation of sediment flux from the Colorado and Nile, rivers that previously each discharged more than 0.1×10^9 t/year. The Rhône carries only about 5% of the load it did in the 19th century,[12] and the Indus River discharges less than

20% of the load it did before construction of barrages in the late 1940s.[13] Since 1950, the number of dams in the world has increased more than sevenfold; in China alone, the number between 1950 and 1982 increased from 8 to 18,595 (van der Leeden et al.,[14] their Table 8-5)! Given the fact that 95% of the dams in the Americas are located in the United States, Canada, Mexico, and Brazil, and that 99% of the dams in Asia are in China, India, Japan, and South Korea, the potential for future dam construction in other countries is obvious.

Stemming river discharge by damming or diversion can have a major impact upon the coastal zone as well as the river system. Decreased or diverted river flow can lead to increased saltwater intrusion into coastal groundwaters, and water-logging, salinization, and desertification can accentuate the need for drilling shallow and deep wells for ground water, which in turn can lead to accelerated subsidence (see below). Decreased freshwater discharge means diminished nutrient flux to estuarine and coastal waters, which adversely affects biological productivity.

The decreased flux of river-borne sediment also can result in increased river channel erosion (and thus endanger supports for bridges and other structures) and shoreline erosion. The effects, however, may not be immediately felt. For example, in 1855, the Huanghe's course shifted abruptly to the north into the Gulf of Bohai, meaning that an annual sediment influx of 1.1 billion tons to the delta front on the Jiangsu coast virtually ceased. Yet, except for local erosion around the abandoned river mouth, the Jiangsu coast continued to prograde until after WW II, after which the coast has been eroding rapidly (average loss of 68 km^2/year). Comparison of bathymetric soundings from the mid-19th century (admittedly few and inaccurate) with recent echo-soundings suggests a general deepening of the submarine delta of 1 to 3 m. Presumably at least a portion of this eroded sediment was transported landward, where it accumulated on the shorefront. Only when the submarine delta reached some new equilibrium did the abandoned delta begin eroding, albeit 90 years after the river mouth had shifted.

There are two other impacts of river diversion, however, that are perhaps not as well documented in the general literature; these are the effects on subsiding deltas and mangrove forests. In natural conditions, delta subsidence is compensated by the deposition of flood-derived sediment. Some low-lying areas experience natural subsidence rates as great as 1 to 10 cm/year, the result of regional and local tectonic subsidence as well consolidation and dewatering of underlying sedimentary se-quences.[15] Where the sediment load is prevented from overflowing onto the subsid-ing delta, often the result of levees, diversion channels, or damming, subsidence may not be compensated by sediment accumulation. An example of the resulting "coastal retreat" is given in the discussion of the Mississippi Delta in a following section.

Mangroves are one of the most important but also most fragile coastal ecosys-tems. In addition to serving as a luxuriant source of biodiversity, mangroves are a valuable environmental and economic resource, acting as nurseries for many commercially valuable fish and shellfish, and sources of wood and food for local inhabitants. Moreover, the mangrove community acts as a sediment trap, thereby retarding coastal erosion.[16] However, their range of tolerance is very narrow, as many of the mangrove species require brackish waters and relatively low rates of sedimentation and subsidence in which to colonize and grow. Declining health of mangrove forests therefore can have particularly deleterious effects on tropical deltas.[17] Damming of rivers means that the brackish water conditions needed for the growth of some critical mangrove species no longer exist. Unfortunately, mangroves are under even more immediate pressures, particularly from logging, such that their ability to survive the next century must be considered problematic.[16] The Philippines, for example, has only about 25% of its original mangrove forests still remaining.[18]

RIVER DIVERSIONS AND THEIR EFFECTS: SOME EXAMPLES

The number of rivers that have been regulated in recent years has increased dramatically. As of 1990, an estimated 13% of the global river flow was trapped within artificial reservoirs.[19] The economic benefits of these diversions cannot be questioned; in some cases they may have saved regions or entire countries from the effects of severe floods or (ironically) droughts. But there are also environmental and economic impacts that result from these dams that should not be minimized.

In the following paragraphs we discuss some major rivers that have been affected by dams and other diversions this century, as well as rivers that will be increasingly dammed in future years. These rivers can serve as examples of what may happen in the next century as many river basins are increasingly impacted by both climatic change and anthropogenic activities.

The Indus River (Pakistan)

The Indus River, draining from one of the cradles of early mankind, is amongst the largest rivers in terms of drainage area (970,000 km^2) and previously also in terms of water discharge and sediment load. With its headwaters at elevations greater than 4000 m, the river drains the arid to semi-arid Himalaya Mountains along northern Pakistan, Afganistan, and western India. Upon leaving the mountains at Attock, the river traverses the 1000-km-wide Indus Plain to the Indian Ocean just east of Karachi (Figure 4–4). River flow is highly dependent upon the monsoonal climate. More than 80% of the water flow occurs during the six months between May and October.[20] Prior to widespread damming, mean runoff was only 245 mm/year, reflecting the relatively low net precipitation in the drainage basin.

Historically, the Indus has carried a heavy sediment load, the result of the steep, mountainous terrain, sparse vegetation (aided by poor agricultural practices), and the periodically intense monsoon-related rains. The annual sediment load measured at Mandari (near the edge of the broad Indus Plain) between 1960 and 1972 averaged about 300 Mt, but at Sehwan, about 700 km from the mountains and 300 km from the Arabian Sea, the sediment load during the same period was generally less than 150 to 200 Mt. During years with unusually heavy rains, however, the sediment load of the river can be substantially higher than these values (see Figure 4-5).

Indus water and sediment discharge have been markedly influenced over the past 50 years by the construction and utilization of embankments, barrages, and dams.

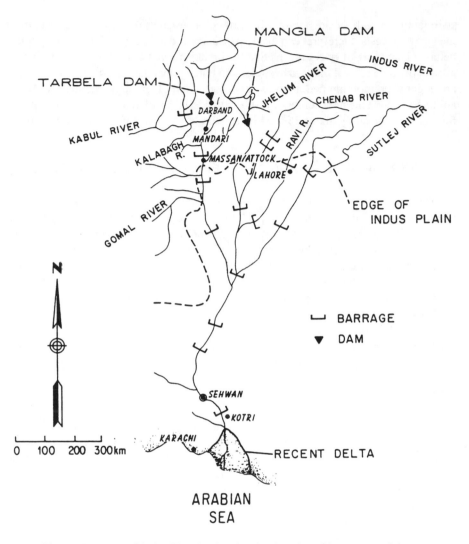

Figure 4–4 Map of Indus River basin, showing location of barrages and dams.

Interestingly, each type of project has had a particular impact on Indus flow. By preventing bank overflow, embankments and dikes, in use since British occupation of Pakistan, have resulted in an increased salinity of the Indus Plain groundwaters and formation of salt crusts that locally have reduced the agricultural utilization of the soils.[20] Channels and barrages have been particuarly useful in irrigating large sections of the Indus Plain. As a result, however, large quantities of soil settle out from the river. At Kotri, only about 50 km from the river mouth, for instance, the sediment discharge dropped from an average of 250 Mt/year to less than 50 Mt/year with the widespread construction of barrages in the mid-1940s. Interestingly, however, because most of the

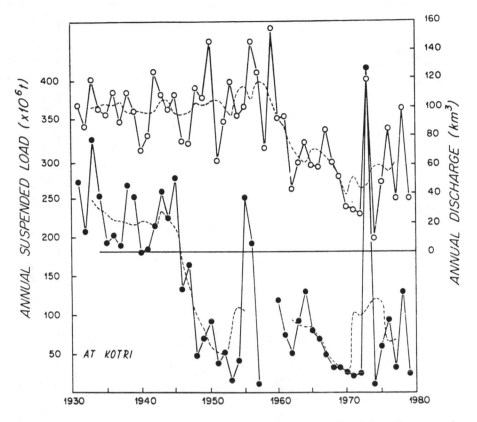

Figure 4–5 Discharge and sediment loads of the Indus River at Kotri (see Figure 4-4 for location) for the years 1931 to 1979. Barrage activity accelerated in the late 1940s; major dam construction was in the early 1960s.

diverted water ultimately flowed back into the main river branch, water discharge at Kotri showed little or no effect of this barrage construction (Figure 4-5). What did affect the water flow was the construction of a series of major mountain dams in the early 1960s, with average flow falling by more than 50%. Even with the installation of dams and barrages, however, major storms still result in water and sediment discharge at or exceeding former levels. One such event occurred in 1973 (Figure 4-5), and it is likely that the heavy floods of 1992 had similar impact. Whatever the effect of these high-energy events, the fact remains that water and sediment discharge from the Indus are about 50 and 20%, respectively, of levels 50 years ago.

Several effects of this river diversion have been noted, such as the salinization of groundwaters and soils. In addition, the mangrove forests in the Indus Delta have been largely destroyed, the result of the increased estuarine salinities.[17] Widespread erosion, the result of decreased sediment flux to the northern Arabian Sea, has not been reported (as far at these writers are aware), but ultimately the unpopulated Sind coast to the east may experience shoreline retreat.

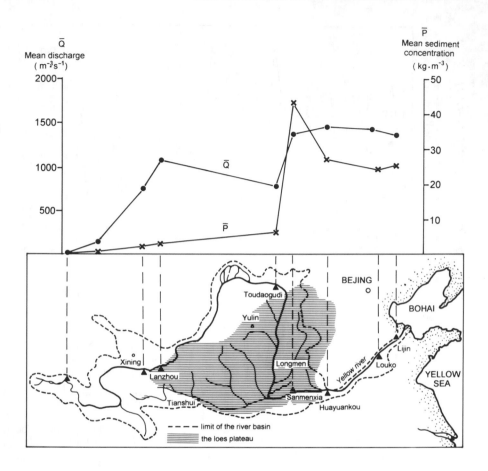

Figure 4–6 Water discharge (Q) and sediment concentration (P) at selected stations on the Yellow River, 30-year average, 1950 to 1979. Note the abrupt increase of sediment concentration after the river flows through the loess plateau. (Modified after Ren and Shi.[21])

Monsoonal Rivers in China — the Huanghe and Changjiang

The Yellow River (Huanghe) and Yangtze River (Changjiang) are often cited as examples of midlatitude monsoonal rivers. However, there are considerable differences in the hydrology of the two rivers. The Huanghe drains semiarid and semi-humid northern China, with a mean annual precipitation of 400 mm throughout the drainage basin; the upper reaches of the river flows through an arid region (Figure 4-6) where annual precipitation is less than 250 mm. The defining aspect of the Huanghe is the fact that the middle reaches flow through the loess plateau from which it obtains its very heavy sediment load; average sediment concentration exceeds 20 g/l, about 100 times greater than most rivers. As the river flow is insignificant (compared to the Changjiang) and the rainfall variable, human intervention on the

Huanghe has been considerable and the river should be particularly susceptible to future climatic change and human activities.

The flow of the Huanghe varies greatly in both time and space. The highest flow occurs at Huayuankou station, about 780 km from the sea, decreasing seaward because of the lack of tributaries and the loss of water via seepage in this "hanging river", where the river bed is 5 to 10 m above the adjacent flood plain.[21] Annual variation in flow is best depicted in the 1920 to 1958 record at Shaanxian, when there was relatively little disturbance by human activity. During this period, mean annual flow was $425 \times 10^8 m^3$, but during 1920 to 1932, mean annual flow was 21% less than the longer term mean, whereas during 1933 to 1958, flow was 10.7% higher than the mean. If we connect the Shaanxian data with those at Sanmenxia station, which replaced it in 1959, we see fluctuating periods of high and low flow between 1960 and 1981, with a more marked low water interval between 1970 and 1975. The last 10 years have been one of drought in northern China. Coupled with increasing withdrawal of river water for various uses, the flow has decreased significantly. At Lijin, annual flow from 1980 to 1987 was only $303 \times 10^8 m^3$, and in 1987, the driest year, it was only $109 \times 10^8 m^3$, about 25% of the long-term average (Figure 4-7).

The Huanghe is one of the most heavily sedimented rivers in the world, with a mean annual load of 1.6×10^9 t at Shaanxian and 1.1×10^9 t downstream at Lijin. Variation in annual sediment load is much greater than is annual water flow. For example, between 1950 and 1987, the ratio between highest and lowest annual load was about 20, whereas the ratio of highest and lowest annual flows was only 6. Just as the flow has decreased in recent years, the load has fallen — at Lijin the average annual sediment load between 1980 and 1987 was 0.62×10^9 t. The sediment load is not simply a factor of annual runoff, but also distribution and intensity of rainfall. At Suide, in northern Shaanxi, as the rainfall increases in intensity from 1:2.5:6, the ratio of soil erosion increases 1:37:233.[22] As a result, some years are characterized by weak discharge but heavy loads, and some years the opposite, as seen in the comparison of discharge, sediment load, and suspended matter concentration (Figure 4-7).

The climate and hydrologic regime of the Changjiang result in a water discharge (9100×10^8 m³/year) more than 20 times that of the Huanghe. In contrast to the Huanghe basin, the Changjiang drains a humid region, with a net annual precipitation greater than 1000 mm, and locally greater than 2000 mm. Moreover, because the Changjiang drainage basin covers a variety of climatic regimes and because of the regulating effect of the large lakes in the middle reaches of the river (Figure 4-8), variation in annual flow is much smaller than that of the Huanghe[23] (Figure 4-9). Similarly, the modulating effect of the large basin area, fifth largest in the world, means that the sediment load (about 0.5×109 t/year) varies only slightly from year to year. Still, in terms of human impact, the Changjiang is one of the most heavily impacted in the world, as nearly 500 million people live within its drainage basin.

There is great uncertainty on the nature and magnitude of climatic change in East Asia over the next century, but recent climatic modeling in China indicates that northern China will experience climatic warming and a reduction in soil moisture.

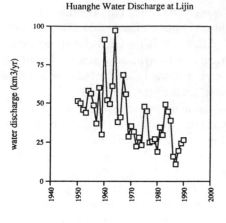

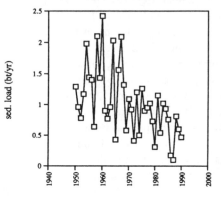

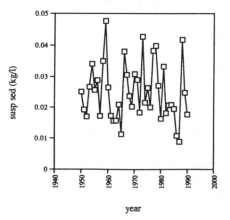

Figure 4–7 Annual water discharge, sediment load, and suspended sediment concentration at Lijin (see Figure 4-6 for location). Note that although flow and load have decreased markedly in recent years, average suspended sediment concentration has continued to range between 20 and 40 g/l.

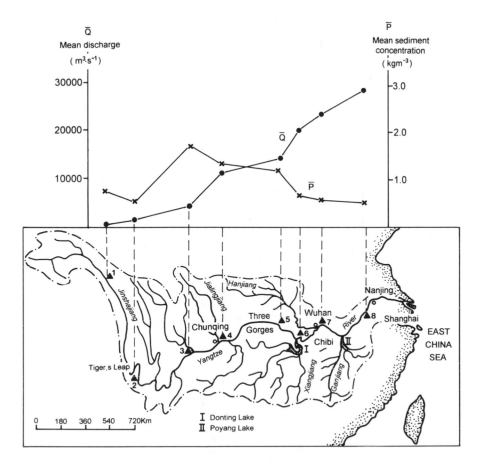

Figure 4–8 Changjiang drainage basin, showing major hydrologic stations and their mean water discharge and suspended sediment concentration.

Even if precipitation remains constant, however, evapotranspiration and evaporation will increase as air temperatures become warmer. Thus, on a climatological basis alone, flow of the Huanghe may decrease in the next century. However, a number of large-scale human activities could complicate the river flow much more seriously. Current economic plans in the upper Huanghe valley call for an additional utilization of 230×10^8 m³ of water for water-consuming industries as well as irrigation projects (Figure 4-10). In the lower valley, the rapidly growing demand for water in agriculture and urban activities decreased the natural flow to the sea by nearly one third in the 1970s compared to the 1950s, and it is estimated that flow at Lijin by 2000 will be reduced by an additional 16% relative to 1980 to 1987.

On the other hand, a proposed interbasin water transfer project is being designed to transfer water from the Changjiang to the Huanghe basin. When completed, a western route will divert an estimated 200×10^8 m³ of Changjiang water to the

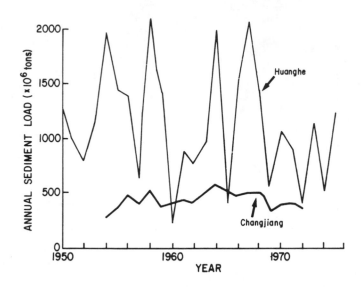

Figure 4–9 Average sediment loads of the Huanghe and Changjiang for the years 1950 to 1978. Note that the flux of sediment from the Changjiang is very small compared to the marked annual variations seen in the Huanghe.

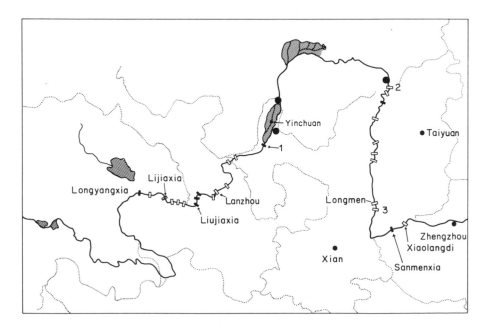

Figure 4–10 Major reservoirs along the Huanghe, showing those completed (■), under construction (▨) (as of 1992), and planned (□).

water-deficient Huanghe. Because of the difficulty of this project, construction will not be completed until the next century, although diversion from the Yalongjiang, easiest in terms of engineering, will add 45×10^8 m^3 yearly. In total, the completion of the western route probably will counter-balance the increased need for water in the Huanghe valley. Considering all these factors, it seems likely that Huanghe flow to the sea in the next century may only be half that of the 1950 to 1979 long-term average. In terms of sediment discharge, a reduction in river flow may result in a corresponding decrease in sediment discharge, although this assumes a continued improvement in soil conservation in the loess plateau. Preliminary study indicates that sediment transfer to the sea may be reduced to 0.4 to 0.5×10^9 tons/year.[24]

Because it drains a very large basin, much of which is humid, the Changjiang should be less susceptible to future change. Being located at a lower latitude, the magnitude of climatic warming in the Changjiang drainage basin over the next century probably will be smaller that in the Huanghe basin. Even if precipitation decreases and evaporation increases, flow may remain near present-day levels because of melting Tibetian and Sichuan glaciers, and the possible increase in tropical cyclones.

On the other hand, human activities already are reducing river flow. The interbasin transfer to the Huanghe ultimately will have the greatest effect, and already about 500 m^3/s is being diverted from the lower Changjiang to northern China. By the middle of the 21st century, when the entire project is completed, approximately 9% of the Changjiang's present flow will have been diverted. Moreover, the present-day modulation of flow by the large midreach lakes may be lost within the next 50 years if the lakes continue to shrink at current rates. If this happens, yearly variation in river flow (even with the expected regulation of flow by the Three Rivers Gorge Dam) will increase.

Construction of a series of large reservoirs in the river basin will trap a considerable percentage of the river's sediment. In particular, the recently completed Three Rivers Gorge Dam will create a reservoir with a capacity of 39×10^9 m^3. As the dam controls 0.51×10^9 t of sediment annually, more than the total annual sediment discharge at the river mouth, the effect of this dam and other large reservoirs on the coastal dynamics at and near the Changjiang's mouth may be considerable, particularly given the tremendous human impact that may result from these changes.

Mississippi (United States)

The Mississippi River is the largest river system in North America, draining an area of 3.3×10^6 km^2. The second largest river system, the Mackenzie (Canada), is less than 60% as large (1.8×10^6 km^2). In terms of both water and sediment discharge, the Mississippi also ranks as the largest in the United States (about 500 km^3 and 230×10^9 t/year, respectively). In terms of water discharge, the Mississippi and Ohio river branches are the major contributors, whereas the Missouri is the major sediment supplier to the system.

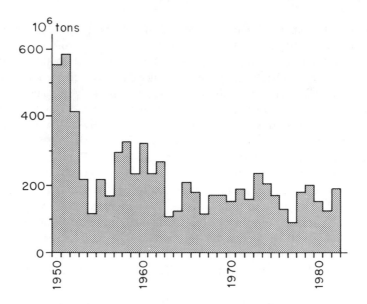

Figure 4–11   Annual sediment load (in millions of English tons) for the Mississippi River at Baton Rouge (southern Louisiana) betwen 1950 and 1982. After Meade and Parker.[34]

The Mississippi discharges into the Gulf of Mexico through a birdsfoot delta south of New Orleans, Louisiana. Throughout the Holocene, however, the river mouth has tended to migrate over a 300-km wide swath. Were it not for the efforts of the U.S. Army Corps of Engineers, in fact, the river probably would have diverted naturally into the Atchafalaya River some years ago. As it is, the Atchafalaya carries approximately 30% of the Mississippi's annual discharge.

Because of the historically important role the Mississippi drainage basin has played in the agriculture of the United States and in terms of transport to and from the central parts of the country, the Mississippi has been heavily affected by human activity. Despite the heavy erosion of cultivated land in the basin (see Meade et al.,[25] their Figure 9B), the discharge of sediment from the Mississippi probably is only half of what it was prior to European colonization in the late 18th century. Although farming activity in such places as the Ohio River Valley led to increased sediment flux to the Mississippi system, according to R. H. Meade (oral communication) buffalo grazing was sufficiently intense and prairie fires were sufficiently common that natural erosion levels also were high.

Moreover, the many dams and channel-stabilization works throughout the river basin have detoured and stored half the sediment. Meade et al.[25] estimate that in 1700 the Mississippi's load was about 400 Mt/year. The 1951 to 1965 average given by Holeman[5] was 310 Mt, but this value is misleading, since the completion of a series of dams on the Missouri in the mid-1950s decreased sediment discharge (at Baton Rouge) from more than 500 Mt for the years 1950 to 1952 to less than 300 Mt in 1960 to 1962, to about 200 Mt after 1962 (Figure 4-11).

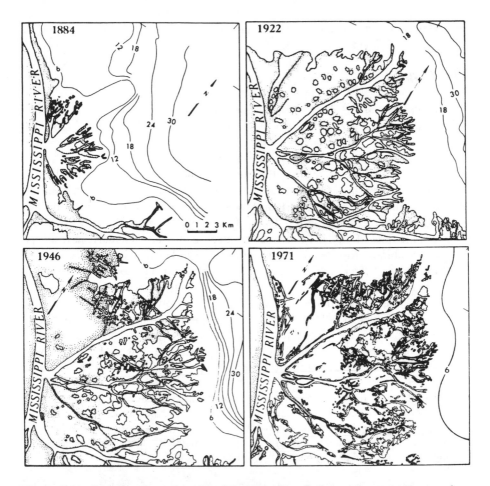

Figure 4–12 Evolution of a portion of the Mississippi River Delta over the past 100 years, from
the growth of a subdelta in the late 19th century and early 20th century to a
marked land loss in the latter half of the 20th century. Note that most of the land
loss occurs in the interior of the delta, not along the shoreface, suggesting that this
is the result of uncompensated subsidence. Since 1950, the sediment discharge
of the Mississippi past Baton Rouge has fallen by half (see Figure 4-11), and
newly constructed levees probably have restricted flood-related spillover even
more. (From Wells, J. T. and Coleman, J. M., 1987. *Estuarine Coastal Shelf Sci.*,
25, 111–125. With permission.)

One result of the decreased sediment load has been the loss of coastal lands in the
state of Louisiana. However, in contrast to the commonly perceived coastal erosion,
much of this land loss has been the result of the widening of coastal lakes and the
drowning of coastal marshes (Figure 4-12). The cutting off of flood-derived sediment
means that the subsiding coastal low lands do not receive the compensating sediment
input from annual floods. As land management and river entrainment increase,
coastal retreat probably will continue, perhaps even increase. The future stability of
the Mississippi Delta is even more questionable if and when its path shifts to the

Achafalaya River, which at present has been stabilized only by the efforts of the U.S. Army Corps of Engineers.

Nile (Egypt)

Although the Nile Delta only occupies about 2% of Egypt, it accounts for nearly half of the cultivated land and houses about half the country's population. Moreover, historically the Nile River annually supplied nutrient-rich soils to the flood plains and nutrient-rich waters to the coastal and offshore areas.

In terms of both basin area (3×10^6 km^2) and water discharge to the Mediterranean (generally less than 60 km^3/year), the Nile River is one the largest rivers in Africa. Draining much of arid eastern Africa, however, its runoff historically has averaged less than 30 mm/year, less than half that of the Huanghe. Moreover, like the Huanghe, the flow of the Nile fluctuates both seasonally and annually, in response to sporadic rainfall. Prior to completion of the Aswan Dam, sediment discharge of the Nile was more than 120 Mt, ranking it among the world's ten largest rivers.

Portions of the Nile's flow have been diverted for agricultural purposes for many hundreds of years, and with completion of the Aswan Low Dam in the early part of this century, as much as 60% of the river's flow was lost to irrigation, evaporation, and seepage.[26] Due to the limited storage capacity of this first Aswan Dam, however, as well as the increased need for hydroelectric power, the Aswan High Dam, was built in the early 1960s. According to Sestini[26] (pp. 576–577),

> The High Dam provides a controlled mean annual discharge of 58.6 km^3 of water downstream, which has enabled expansion of the cultivated area by 1.3 feddans and a conversion of 700,000 feddans from basin to perennial irrigation. About 10,000 MW of electricity are generated annually, navigation conditions along the Nile have improved, and flood protection is now guaranteed.

While it is doubtful if Egypt could have successfully survived the drought of the late 1970s and early 1980s without the storage capacity of Lake Nasser (160 km^3), the Aswan High Dam clearly has had major environmental impacts downstream, particularly along the delta front. The lack of nutrient-rich waters from the Nile (most of the river water that reaches the lower Nile is utilized for agricultural and domestic/industrial purposes; only 2 to 3×10^6 km^3 is discharged presently into the Mediterranean) has meant a drastic reduction in coastal and offshore fisheries. The year after the completion of the Aswan High Dam, for instance, the sardine fisheries decreased by 95%;[27] increased fish catches in the late 1970s reflected increased effort and efficiency rather than increased productivity.[28] The lack of freshwater influx also has meant increasing salinization of the soils and the corresponding need for increased drilling of deep wells for groundwater.[29] Removal of these waters may well increase the subsidence rate of the lower delta, already in places as much as 0.5 cm/year.[30]

Finally, the sediment flux to the Mediterranean from the Nile has almost ceased, resulting in rapid erosion along the now mostly inactive Damietta and Rosetta mouths

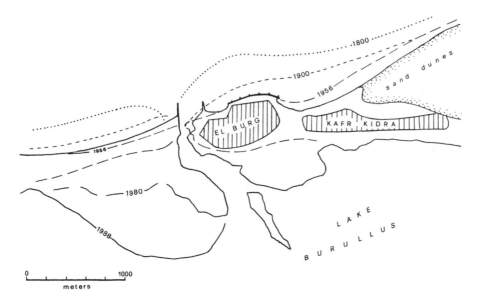

Figure 4–13 Erosion of the coastal area separating Lake Burullus on the Nile Delta from the adjacent Mediterranean. In part erosion of this coastline was the result of decreased discharge from the Nile River in the late 19th and early 20th centuries; increased erosion in the 1980s presumably reflects the influence of the Aswan Dam. (From Sestini, G., 1992. In L. Jeftic, J. D. Milliman, and G. Sestini, Eds., *Implications of Climatic Changes in the Mediterranean Sea*. Pergamon Press, pp. 535–601. With permission.)

of the delta. In fact, due to a prolonged decrease in rainfall in Ethiopia in the late 19th century and activation of the lower Aswan Dam in the early 1900s, erosion has been noticeable along much of the Nile Delta for the past 100 years (Figure 4-13). With the construction of the Aswan High Dam, however, erosion has accelerated, and locally it has exceeded 100 m/year, although, interestingly, erosion mostly has been confined to promontories (Figure 4-14). As a whole, the delta continues to aggrade,[31] perhaps in response to onshore transport of sediment.[32] At some point, when a new equilibrium is reached, coastal erosion probably will accelerate, perhaps leading to the breaching of the narrow coastal dunes that presently confine the highly productive coastal lagoons (i.e., Lakes Maryut, Idku, Burullos, and Manzalah) (Figure 4-13).

GLOBAL IMPACT OF FUTURE MODIFICATIONS AND CLIMATE CHANGE

Rivers and adjacent coastal areas are under increasing environmental stress at a time when engineering projects and possible climatic change from greenhouse warming threaten even greater stress. The coastal zone has far greater impact on human activities than any other marine environment, but the reverse is also true. The management of the coastal zone, therefore, must take into account the effect of

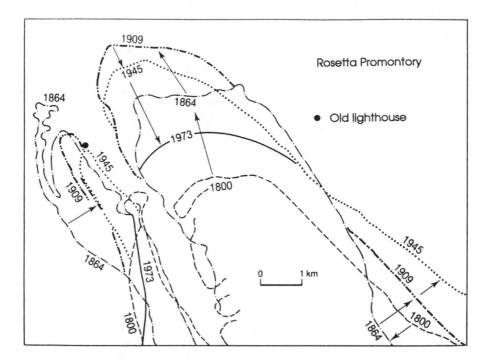

Figure 4–14 Evolution of the Rosetta Promontory (Nile Delta) since 1800. Note that the promontory prograded until the early part of this century, after which it has eroded at an accelerated pace. Annual shoreline loss at the promontory itself was 75 m between 1945 and 1973, and assuming that erosion before 1964 was similar to that between 1909 and 1945, erosion after 1964 must have been greater than 175 m/year!

Years	Average accretion/erosion (m/year)
1800–1864	+31
1884–1909	+27
1909–1945	−17
1945–1973	−75

(After Sestini, G., 1992. *Implications of Climatic Changes in the Mediterranean Sea*, Jeftic, L., Milliman, J. D., and Sestini, G., Eds., Pergamon Press, 535–601. With permission.)

human activities as well as impacts resulting from corresponding changes in the coastal zone.

With the projected climatic warming during the next century, higher rates of evapotranspiration and (locally) altered rainfall could lead to higher rates of erosion and thus greater input of fluvial sediment to the sea.[33] As developing countries (wherein lie most of the major rivers in terms of discharge and sediment load — see Table 1) continue their economic development, however, there will be increased demand for "river control and utilization," i.e., diversion projects in the form of embankments, dams, and barrages. Accelerated need for increased hydroelectric power, maximizing agricultural use of the waters (by irrigation), and minimizing the impact of floods or droughts are obvious reasons for such projects. Concomittant with these trends, however, are escalating populations along river courses and increased upland erosion rates due to poor agricultural practices and increasing deforestation.

Many of the existing dams throughout the world will need to be replaced within the next 50 years, in large part because the reservoirs behind them will be filled with trapped river sediment. Many North American dams will be approaching 100 years of age by that time (their predicted life expectancy), and many newer dams in Asia will have shorter-than-predicted life expectancies because of high levels of sedimentation in their reservoirs. If sluicing methods akin to those used by the Chinese on their Huanghe and Changjiang dams are utilized, however, the useful life of many dams across sediment-laden rivers may be extended.

If river management is undertaken without regard to the nature of the river as well as the entire river-coastal ecosystem, one can predict a gloomy yet logical sequence of events: increased river diversion; decreased flow (particularly during floods); less overtopping of banks; less sediment accumulation on the (subsiding) delta; increased shoreline erosion and decreased mangrove (and coastal biological) productivity; greater local sea level rise because of eustatic sea level rise and the resulting uncompensated subsidence; downstream waters become increasingly polluted and salt-water intrusion contaminates shallower water supplies; greater need for groundwater as rivers are dammed upstream; accelerated subsidence rates as groundwater removal causes accelerated soil compaction. Using the most pessimistic global and local scenarios, for example, Milliman et al.[29] showed that eustatic sea level rise and local subsidence could result in both Bangladesh and Egypt losing more than 25% of their habitable land by the year 2000, a particularly gloomy prospect for two of the world's most overpopulated countries.

The severity of climate change upon individual river basins in part depends on the way in which the rivers and their basins are utilized. Moreover, the impact of individual or collective river diversion projects depends more upon *how* they are done rather than *if* they are done. Taking the holistic point of view of the river-coastal system, particularly to the river downstream of any particular project, may lead to more imaginative decisions in river basin management. The dam that saves a population from annual flooding ultimately will do little good if it also hastens coastal erosion that imperils that same population. Because the need for river diversion projects is increasing dramatically, these types of decisions will need to be made increasingly in the coming years; moreover, they will require input from a broad spectrum of public, private, and academic perspectives.

REFERENCES

1. Milliman, J. D., 1991. The flux and fate of fluvial sediment and water in coastal seas. In R. F. C. Mantoura, J.-M. Martin, and R. Wollast, Eds., *Ocean Margin Processes in Global Change*. John Wiley and Sons Ltd., pp. 69–89.

2. Meybeck, M., 1979. Concentrations des eux fluviales en éléments majeurs et apprôts en solution aux océans. *Rev. Geol. Dyn. Geogr. Phys.*, 21, 215–246.

3. Baumgartner, A. and Reichel, E., 1975. *The World Water Balance*. Elsevier, Amsterdam.

4. Milliman, J. D. and Syvitski, J. P. M., 1992. Geomorphic/tectonic control of sediment discharge to the ocean: the importance of small mountainous rivers. *J. Geol.*, 100, 525–544.

5. Holeman, J. N., 1968. Sediment yield of major rivers of the world. *Water Resour. Res.*, 4, 737–747.

6. Lisitzin, A. P., 1972, Sedimentation in the world ocean. *Soc. Econ. Paleont. Miner.*, Spec. Publ. 17.

7. Milliman, J. D. and Meade, R. H., 1983. Worldwide delivery of river sediment to the oceans. *J. Geol.*, 91, 1–21.

8. Saunders, I. and Young, A., 1983, Rates of surface processes on slopes, slope retreat and denudation. *Earth Surf. Proc. Landforms*, 8, p. 473–501.

9. Myers, N., 1988. Tropical deforestation and climate change. *Environ. Conserv.*, 15, 293-298.

10. Arnold, J. E. M., 1987. Deforestation. In D. J. McLaren and B. J. Skinner, Eds., *Resources and World Development*, John Wiley and Sons, Ltd., London, pp. 711–725.

11. Milliman, J. D., Qin, Y. S., and Ren, M. E., 1987. Man's influence on the erosion and transport of sediment by Asian rivers: The Yellow River (Huanghe) example. *J. Geol.*, 95, 751–762.

12. Corre, J. J. et al., 1992. Implication des changements climatiques. Etude de case le Golfe du Lion (France). In J. D. Milliman, Ed., *Climatic Change and the Mediterranean*. Edward Arnold, Seven Oaks, U.K., pp. 328–424.

13. Milliman, J. D., Quaraishee, G. S., and Beg, M. A. A., 1984. Sediment discharge from the Indus River to the ocean: past, present and future. In Haq, B. U. and Milliman, J. D., Eds., *Marine Geology and Oceanography of the Arabian Sea and Coastal Pakistan*. Van Nostrand Reinhold, New York, pp. 65–70.

14. van der Leeden, F., Troise, F. L., and Todd, D. K., 1990. *The Water Encyclopedia*. Lewis Publishers, New York.

15. Dolan, R. and Goodell, H. G., 1986. Sinking cities. *Am. Sci.*, 74, 38–47.

16. Ormond, R., 1988. IOC/Unesco Workshop on Regional Cooperation in Marine Science in the Central Indian Ocean and Adjacent Seas and Gulfs. Workshop Report No. 37, Supplement, pp. 167–193.

17. Snedaker, S. C., 1984. Mangroves: a summary of knowledge with emphasis on Pakistan. In Haq, B. U. and Milliman, J. D., Eds., *Marine Geology and Oceanography of the Arabian Sea and Coastal Pakistan*. Van Nostrand Reinhold, New York, pp. 255–262.

18. Gomez, E. D., 1988. Overview of environmental problems in the East Asian Seas region. *Ambio*, 17, 166–169.

19. Pearce, F., 1991. A dammed fine mess. *New Scientist*, May 4, p. 36–39.

20. Beg, M. A. A., 1977. The Indus River basin and risk assessment of the irrigation system. Int. Working Seminar on Environmental Risk Assessment in an International Context. Tihanyi, Hungary.

21. Ren, M.-e. and Shi, Y. L., 1986. Sediment discharge of the Yellow River (China) and its effect on sedimentation of the Bohai and Yellow Sea. *Cont. Shelf Res.*, 6, 785–810.

22. Zhu, M.-M. and Ren, M.-e., 1991. Research on the course of the formation and renovation of the Loess Plateau in China. *Popul. Res. Environ. China*, 1, 21–38.

23. Shi, Y.-L., Tang, W., and Ren, M.-e., 1985. Hydrological characteristics of the Changjiang and its relation to sediment transport to the sea. *Cont. Shelf Res.*, 4, 5–15.
24. Ren, M.-e. and Zhu, X. M., 1991. Anthropogenic effect on changes of sediment discharge of the Yellow River, China since the Holocene. *Quat. Sci.*, 1992
25. Meade, R. H., Yuzyk, T. R., and Day, T. J., 1990. Movement and storage of sediment in rivers of the United States and Canada. In M. G. Wolman and H. C. Riggs, Eds., *The Geology of North America*, Volume 1, Surface Water Hydrology. Geological Society of America, pp. 255–280.
26. Sestini, G., 1992. Implications of climatic changes for the Nile Delta. In L. Jeftic, J. D. Milliman, and G. Sestini, Eds., *Climatic Change and the Mediterranean*. Edward Arnold Press, Seven Oaks, U.K., pp. 535–601.
27. Wahby, S. D. and Bishara, N. F., 1981. The effect of the River Nile on Mediterranean Water, before and after the construction of the High Dam at Aswan. In *Proceedings of a Review Workshop on River Inputs to Ocean Systems*. United Nations, New York, pp. 311–318.
28. Halim, Y., 1991. The impact of human alterations of the hydrological cycle on ocean margins. In R. F. C. Mantoura, J.-M. Martin, and R. Wollast, Eds., *Ocean Margin Processes in Global Change*. John Wiley and Sons Ltd., pp. 301–327.
29. Milliman, J. D., J. M. Broadus, and F. Gable, 1989. Environmental and economic implications of rising sea level and subsiding deltas: the Nile and Bengal examples. *Ambio*, 18, 340–345.
30. Stanley, D. J., 1988. Subsidence in the northeastern Nile Delta: rapid rates, possible causes and consequences. *Science*, 240, 497–500.
31. Smith, S. E. and Abdel-Kader, A., 1988. Coastal Erosion along the Egyptian Delta. *J. Coastal Res.*, 4, 245–255.
32. Murray, S. P., Coleman, J. M., Roberts, H. H., and Salama, M., 1981. Accelerated currents and sediment transport off Damietta Nile promontory. *Nature*, 293, 51–54.
33. Le Houérou, H. N., 1992. Vegetation and land-use in the Mediterranean Basin by the year 2050: a prospective study. In J. D. Milliman, Ed., *Climatic Change and the Mediterranean*. Edward Arnold Press, Seven Oaks, U.K., pp. 175–232.
34. Meade, R. H. and Parker, R. S., 1985, *Sediment in rivers of the United States*. U.S. Geological Survey Water-Supply Paper 2275, pp. 49–60.
35. Wells, J. T. and Coleman, J. M., 1987. Wetland loss and the subdelta life cycle. *Estuarine Coastal Shelf Sci.*, 25, 111–125.

CHAPTER 5

Response of Estuaries to Climate Change

K. R. Dyer

INTRODUCTION

Estuaries are ephemeral features situated as they are between the land and the sea.[1] They are always prone to change, occurring with a wide range of time scales, some cyclic and others intermittent, those changes originating both on the land and in the sea. The variability is large, and this tends to obscure the slow changes that take place because of the evolution of the natural environment; changes that have been taking place throughout geological time. Estuaries are dynamic and responsive systems. They must be, otherwise they would not have been able to adapt and develop healthy, diverse, and specialized populations of animals and plants. In the end it is they that are the indicators of whether change is acceptable or not. A healthy estuary has a variety of habitats with extensive and specialized flora and fauna. However, many estuaries have been exposed to pollutant input and have become virtually dead. Nevertheless, control of pollutant inputs are restoring some of them, such as the Thames, to acceptable conditions.

It is only in comparatively recent times that estuaries have been valued, and their importance as unique and sensitive ecosystems has been fully appreciated. This has come at a time when mankind has also realized that fundamental alterations are being wrought to the earth, which may cause irreversible changes or accelerate the natural processes beyond the capacity of those systems to adapt. Because of the importance of estuaries, prediction of the possible consequences is needed so that preventive or ameliorative measures can be taken in good time. However, estuaries are complex environments where there are strong coupling and feedback between the physics, sediments, chemistry, and biology. Thus we are still in the situation of describing and quantifying the present processes.

Predicting the consequences of change in the system can benefit from the examination of case histories of the response of estuaries to natural or to man-made change. Unfortunately, we are unable to benefit from the detailed knowledge that could have been available from study of the consequences of industrialization on estuaries. We can record the situation now, but we do not know what it was before the 19th century explosion of port and harbor building, reclamation, and estuary-based industry. Had such information been available, these case studies could have taught us a great deal about the complex interlinking of processes that control the health of the estuarine environment, and the response to drastic alteration. Generally there has been insufficient long-term monitoring to provide the information to quantify the slow changes. We can predict only within the boundaries of the variations that have been measured. To go beyond that is difficult.

Mathematical modeling can incorporate parameterizations of these processes, and can be used as aids for prediction, but such models are particularly sensitive to the boundary conditions, i.e., what the temperatures or the rainfall would be for a particular estuary in the future. Consequently, prediction depends, in the first instance, on an adequate definition of future climate and weather. Following that, conditions in each estuary would have to be separately specified and investigated. These studies are still in their infancy. In this chapter the basic responses of the estuarine system to variation that could result from climate changes are described and illustrated with examples.

ESTUARINE TIME SCALES

Estuaries respond to various influences with a variety of time scales, both internal and external. However, delays and modifications are introduced so that the estuary often acts like an electronic filter, changing the amplitude and the phase of the input signal, with the output into the coastal sea having a different magnitude and timing. Determining the characteristics of the circuit, the processes that control the impedances, resistances, and capacitances is the challenge of estuarine research.

The range of physical time scales operating in estuaries is shown in Figure 5-1. The magnitudes of the various variables are likely to change from one estuary to another. Also, the spectrum of frequencies of each variable will alter both between estuaries and with time. Nevertheless, each estuary could experience the full range of inputs.

Regular cyclic variations are produced by the semidiurnal tidal variations in sea level that flood and expose the intertidal mudflats and marshes. There are spring and neap cycles and the seasonal variations. The associated currents produce turbulence and internal waves, cause mixing, erosion, transport, and deposit sediment, and disperse contaminants.

The dynamic tidal response of an estuary is shown by the delay of the tide between the mouth and the head of the estuary, due to the friction creating a partially progressive tidal wave. This delay is generally of the order of a few tens of minutes,

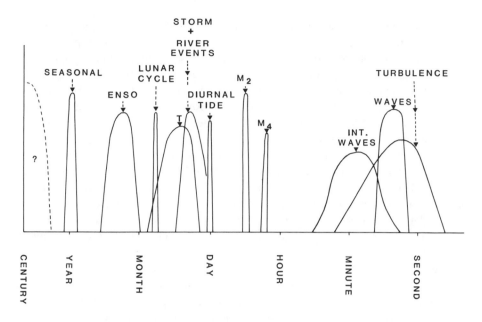

Figure 5–1 Diagramatic representation of the important physical time scales affecting estuaries: T, estuary flushing time; M₂, semidiurnal tidal cycle; M₄, quarterdiurnal tide. The vertical energy scale is unquantified.

but for larger estuaries can be hours. In the tidal Amazon several tidal cycles are progressing at any one time. Consequently, a change in sea level is transmitted through the estuary fairly quickly. Longer period intermittent elevation of the sea level occurs during El Niño-Southern Ocean (ENSO) type events. These can have durations of months up to about a year.

Seasonal variations in river discharge of water and sediment are significant in all estuaries. The response of the salinity in the estuary to changes in river discharge R is given by the flushing time. This is the rate at which fresh water within the estuary is replaced. The flushing time $T = Q/R$, where Q is the fresh water content of the estuary, $Q = (S_s - \overline{S}.V)/S_s$, where S_s is the sea salinity and $\overline{S}$ the mean salinity in the section of the estuary of volume V. The flushing time is generally of the order of a few days to tens of days. Consequently, within a matter of a couple of weeks, the estuary will have responded to a change in river discharge, or a flood event, and achieved a new steady state level.

Intermittent events are produced by storms, river floods, and surges. These characteristically have time scales of the order of 2 to 5 days, which relate to the passage of time for depressions. The storms produce surface waves whose period will depend on wind fetch and water depth.

Wind events cause significant changes to the estuarine circulation, and in order to average their effects, Weisberg has shown that obtaining an estimate of the residual velocity within ±20% of the long-term value would require an averaging time of about 18 days.[2]

For sediment, however, the situation is very different. Estuaries are a trap for fine sediment which can be derived from the land and the sea. The trapping efficiency varies with estuarine type, from effectively zero for salt wedge estuaries and deltas, to 100% for well-mixed estuaries. Since the estuarine circulation is a good sorting mechanism, an estuary can export one grain size, while importing another. Sedimentary time scales may be many hundreds of years. However, within the estuary erosion and deposition cycles can be considerably shorter, with semidiurnal, spring-neap, and seasonal effects exchanging large volumes of sediment.

Many estuaries have not been able to fill with sediments despite ~5000 years of reasonably constant sea level, whereas others have filled to the stage where a long-term equilibrium appears to have been achieved between the sediment inputs and the fluid forces. In these cases there seems to be a balance with the net sediment accumulation equaling the long-term changes in sea level. Thus, the topographic response of the estuary may be thousands of years, depending on the sediment source characteristics.

Chemical response times depend on whether the chemical is dissolved or particulate, in which case it will have the appropriate hydrodynamic or sediment particle response time. However, particularly within the sediment bed, the contaminants can disassociate from the particles, and within the water scavenging from solution onto particles can occur. Scavenging generally occurs within minutes. Disassociation within the bed normally takes days.

Biological response to change will be effective within a few generations, or growing seasons, the creatures first showing signs of stress particularly on reproductive systems, before dying. Obviously, mobile creatures can respond dynamically and are less prone to the consequences of change. Sessile creatures can tolerate change for a few hours by shutting down their metabolic systems. Both intensity and duration of the change are important in determining the ecosystem response.

To be significant, the time scales for the different factors must be compatible. Thus, a chemical change that takes minutes to reach equilibrium will not be directly affected by physical variations of a tidal time cycle, but it is likely to respond to the influences of turbulence. Similarly seasonal variations in river discharge will affect the seasonal biological cycle, but is not likely to directly influence the intensity of the spring bloom. However, in many of the interactions we will be concerned with the subtle second-order effects, since they appear to be important.

CLIMATE CHANGE

There has been an increase in the global atmospheric content of carbon dioxide by about 9% since the mid-1970s. Carbon dioxide is a 'greenhouse gas' in that it allows incoming short-wavelength energy to reach the earth from the sun, but impedes the back radiation from the earth at longer wavelengths. This has led to a temperature increase of 4°C and the direct heating of the ocean has lead to an increase in sea level of 10 cm over the last century. It is predicted that there will be a further rise of 1.5 to 4.5°C by the year 2050 and sea level is predicted to rise 50 cm by 2050, and 1 m by 2100 (Chapter 4).

In addition, the warming is likely to change the weather patterns, and, in particular, increase the area of sea with temperatures greater than 29°C, within which tropical cyclones are generated. This, and the greater temperature gradients between the tropics and the poles, point to an increased storminess and shorter duration of calms. Thus, higher strength and more frequent gales are likely at high latitudes in the winter, with increased rainfall and river discharge (Chapter 3). In the summer warm, dry conditions are predicted.

At lower latitudes movement of the climatic zones is likely to increase the areas of desert and summer drought, as outlined in Chapter 1.

Consequently, for estuaries we have to consider a wide range of changing climatic variables:

1. River discharge of water and sediment
2. Temperature rise
3. Rainfall, magnitude and frequency, both increase and decrease
4. Wind, including surge frequency and wave climate
5. Mean sea level rise
6. Tidal range

RIVER DISCHARGE

Reduced river flow alters the circulation of the estuary, the mixing, and its capacity to dilute, transform, and flush contaminants. River discharge, together with tides, are the two main factors that drive the physical processes in estuaries. Highly stratified estuaries occur when river discharge dominates the tidal motion, and when the mixing is limited to the halocline separating the surface fresh and bottom salty water. For partially mixed estuaries, tidal influence is greater, turbulent mixing is also created by tidal friction at the sea bed, and vertical gravitational circulation is produced.[3] Well-mixed estuaries are dominated by tidal action. Consequently, the degree of salinity stratification, the intensity of the mixing, and the vertical gravitational circulation depend on river discharge and tidal range.

These effects have been formalized in classification schemes.[4-6] Their diagrams can be used to define the changes likely with a change in river discharge. An increase in discharge pushes the salinity down-estuary, steepens the horizontal salinity gradients, increases the stratification, and decreases the intensity of the vertical mixing. Consequently, a partially mixed estuary would become highly stratified. Conversely, decreased river flow would lead to a better mixed system (Figure 5-2).

Empirical relationships can also be established. Bowden and Sharaf el Din,[7] for example, found that the best correlation between salinities and river flow occurred when they used the average river flow for the week before sampling. They established linear relationships between the mean salinity at two sections, and the stratification,

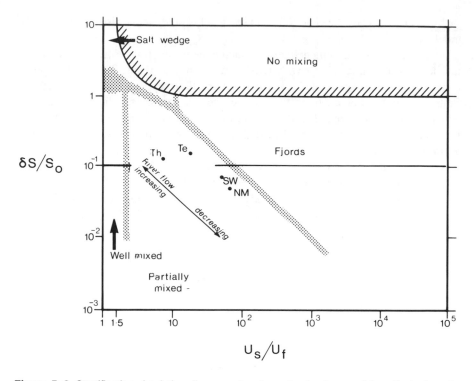

Figure 5–2 Stratification-circulation diagram, showing estuarine type and the effect of varying river flow: δS, surface-to-bottom mean salinity difference, S_o, depth and tidal mean salinity; u_s, surface mean velocity; u_f, freshwater discharge velocity (Q/A); Th, Thames; Te, Tees Estuary; SW, Southampton Water; NM, Narrows of the Mersey. (After Hansen and Rattray.[4])

with river discharge and tidal range. The horizontal and vertical gradients of salinity depended more on river flow than tidal range. However, the predictive capability of this approach is limited by the range of conditions sampled.

The coefficient of longitudal dispersion K_x is useful in prediction of the intrusion length and of the dispersion of pollutants. This is defined as $K_x = R\bar{s}/(A\delta s/\delta x)$, where R is the river flow, A the cross-sectional area, $\bar{s}$ the cross-sectional mean salinity, and $\delta s/\delta x$ the longitudinal gradient. Since, as shown above, $\delta s/\delta x$ is a function of R, the relationship between K_x and R is nonlinear.

Measurement of K_x for the Tay Estuary is shown in Figure 5-3.[8] Near the mouth the salinity gradient is relatively insensitive to river discharge, but further upstream the changing salinity gradient causes the value of K_x to reduce with discharge, above a critical discharge magnitude.

River discharge variation will affect the flushing time of the estuary. This is inversely proportional to R. The flushing time for Boston Harbor (Figure 5-4) shows that the freshwater volume accumulated in the estuary is not constant, but varies with river discharge.[9] There are simple methods for calculating the flushing time and

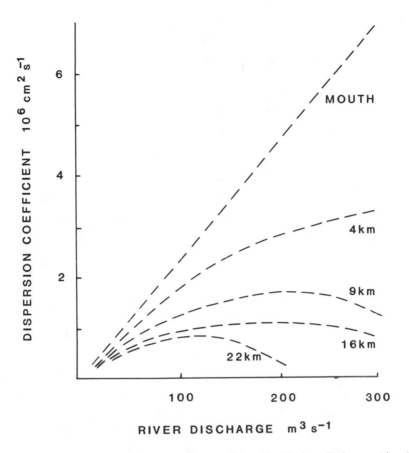

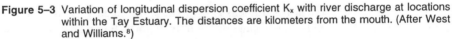

Figure 5–3 Variation of longitudinal dispersion coefficient K_x with river discharge at locations within the Tay Estuary. The distances are kilometers from the mouth. (After West and Williams.[8])

salinity distribution.[10,11] Computer modeling of these effects can give reliable predictions.

There are natural long-period variations in river discharge, e.g., 11 to 15 years, which give variations up to 25%.[12] Consequently, providing the change is within these bounds, prediction of the effects should be possible. However, it is becoming apparent that most rivers have had steadily decreasing river flow for many decades, caused by damming, irrigation, and industrial and domestic water use. The estuaries have changed too. Several examples have been documented by Halim.[13] In San Francisco Bay, river flow has been reduced by more than 50%, and this has led to an increase in the salinity of the bay by 20 to 27%. Reductions of flow in the rivers of the Black Sea and of the Nile have led to drastic salinity changes, with consequential reduction in primary production and fisheries.

The Santee-Cooper River system has been affected by flow diversion.[14] The Santee flow was diverted into the Cooper River in 1941. This caused discharge to

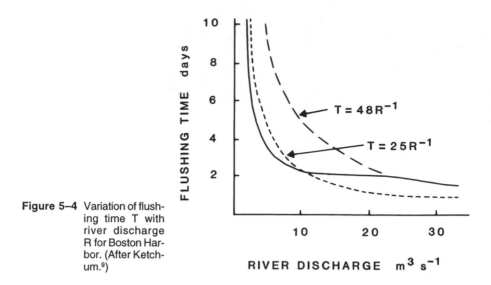

Figure 5–4 Variation of flushing time T with river discharge R for Boston Harbor. (After Ketchum.[9])

change from 2 to 442 m^3 s^{-1}, and salinity moved downstream, the estuary changed from well mixed to partially mixed, and the gravitational circulation increased, as did shoaling. In 1985 partial rediversion caused flow to reduce from 442 to 130 m^3 s^{-1}. Since then the Cooper is well mixed on spring tides, and in a transition zone salinities vary 16% between springs and neaps, and are 10% higher than previously.

An 80% reduction in Eastmain River flow occurred because of damming.[15] The response is shown in Figure 5-5, and salinity reached a new quasi-steady state over a period of about 40 days. The circulation field reached quasi-equilibrium within 8 days. After the flow reduction the estuary became partially mixed rather than salt wedge.

Increasing flood frequency and intensity will cause the estuary to spend more time responding to rapid flow changes. Nichols reports the response of the Rappahannock to changing river flow during Hurricane Agnes.[16] This is one of few papers that help in understanding the effects of transitory events. More work on this will be necessary for predictive modeling.

In tropical areas increasing periods of low or zero river flow will cause longer intervals of hypersaline conditions, during which the estuary can become dominated by evaporation and a reverse salt wedge can develop.[17] The length of the low salinity gap, which varies in duration and degree of salinity, controls the invasion and expansion of species in semiarid estuaries.[18]

The changing river discharge will alter the amount of sediment brought into the estuaries in a more drastic way, since the sediment transport rate is related to the water flow by a power law with a high exponent. Most sediment discharge occurs on the occasional extreme events. Walling and Webb have shown for one example that 90% of the sediment discharge occurs in 5% of the time.[19] During Hurricane Agnes in 1972 more sediment was brought into the Chesapeake Bay than would have been deposited in 30 or so normal years.[20] Under normal circumstances a fraction of the sediment eroded from the catchment is input into the estuary, the rest being stored

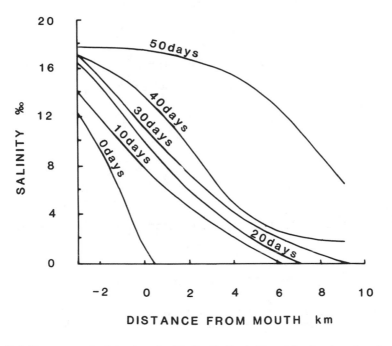

Figure 5–5 The progressive intrusion of salt in the Eastmain River following river diversion. The salinity was measured 2 m below the surface. (After Lepage and Ingram.[15])

in the lower reaches of the river. Phillips reports that in the Waccamaw River/Winyah Bay Estuary system only 4% of the eroded sediment reaches the estuary.[21] Increased magnitude of the flood discharges could potentially remobilize large quantities of this stored sediment. However, increased frequency of flood events will be affected by the phenomenon that when two flows occur close together, the second has a lower sediment concentration.[22] Thus, further increase in variability of sediment discharge will result.

Additionally, the climate change may well affect the rate of weathering, particularly if the agricultural practices change. Damming and control of river discharge also reduces the sediment input into estuaries. These effects are described in Chapter 4. Krone[23] and Nichols et al.[24] have described the consequences of man's intrusion into the catchment of the San Francisco Bay system. This initially led to a major loss of volume because of siltation and reclamation. However, reduced river flow and damming has led subsequently to a decrease in sediment discharge by 60 to 75%.[13]

A major feature of estuaries is the turbidity maximum. This is a zone where suspended sediment concentrations are higher than either in the river or the sea. It is generally located at, or near, the head of the salt intrusion, and is sustained by a combination of tidal pumping and gravitational circulation.[25,26] The turbidity maximum moves dynamically within the estuary with changes in the river discharge. Figure 5-6 shows the movement within the Tay Estuary.[27] The movement of the turbidity maximum within the Tamar estuary, and its relation to the seasonal river discharge cycle has been modeled by Uncles et al.[28] Associated with the turbidity

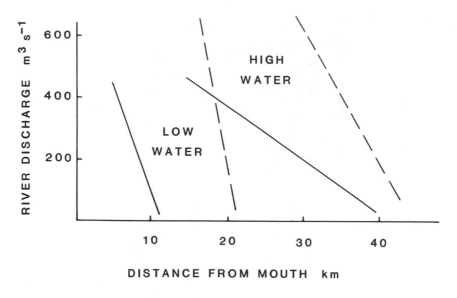

Figure 5–6 Variation of the position of the turbidity maximum at high and low waters with river discharge within the Tay Estuary. (After Dobereiner and McManus.[27])

maximum is a "mud reach", a section of the estuary bed where the shoaling rate is greatest and where the mud is continually eroded and redeposited, exchanging with that in suspension. A permanent shift in the locus of movement of the maximum would produce a change in the sediment distribution, and the location of maximum dredging requirement. Such a shift has occurred in the Seine Estuary, but in this instance produced by reclamation of the mudflats rather than a change in river discharge.[29]

The trajectory of movement of the turbidity maximum at high river flow may reach the mouth of the estuary, in which case direct export of material to the coastal zone can occur. If this becomes a more frequent occurrence, then the trapping efficiency of the estuary will be reduced. The trapping efficiency of the estuary is the ratio of the fluvial sediment input to that accumulated in the estuary. For partially mixed estuaries it can reach 100%.[30] Since a feature of partially mixed estuaries is that the landward residual bottom flow draws in sediment from beyond the estuary mouth, the fluvial sediment is only part of that accumulating. Some of that drawn in is likely to be fluvial material exported at higher river flow stages, but much will be of coastal or marine origin. Additionally, in well-mixed estuaries tidal pumping becomes significant in transporting sediment up-estuary into the turbidity maximum. A filter efficiency can be defined that takes account of the additional accumulation of marine derivation.[31] Figure 5-7 shows diagramatically the filter efficiency of estuaries. The magnitude for well-mixed estuaries is likely to depend on the friction characteristics and on the tidal pumping. Changing river discharge will affect the gravitational circulation and thus the efficiency of trapping. Because of increasing sea level and storminess, increased coast erosion would make more material available

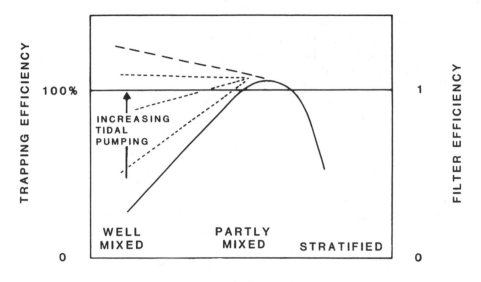

ESTUARINE TYPE

Figure 5–7 Variation of the trapping efficiency (full line) and filter efficiency (dashed lines) of estuaries with estuary type. Increasing tidal pumping increases the trapping action in well-mixed estuaries. (After Nichols[30] and Schubel and Carter.[31])

to silt up estuaries if the new river discharge regime creates a partially or well-mixed estuary.

On the other hand, the 90% flow reduction in the Eastmain River, previously mentioned, led to the development of a turbidity maximum and retention of suspended sediment. Though sediment discharge to the estuary was reduced by a factor of ten, the rates of sediment transport out of the estuary are at least 25 times smaller than before.[32] Similarly, reduction of water flow in the Nile has reduced the sediment load of about 160×10^6 t/year to virtually zero. As a consequence the delta front is retreating at a rate of 100 m/year.[13]

TEMPERATURE

An increase in temperature would decrease the surface water density and create an increased tendency to stratification. One effect of this could be an increased occurrence of anaerobic conditions in the lower layer, if it becomes quasi-stagnant. Reoxygenation occurs at spring tides, and the strength of the vertical mixing affects the severity of hypoxia and the location of the dissolved oxygen (DO) minimum.[33] Generally, however, the effects of salinity changes are likely to outweigh those of temperature on the main water body. Nevertheless, local effects may be large.

During a sunny day maximum heating of the water occurs during the covering of the intertidal mudflats. The greatest effect is thus likely when high tide occurs at midafternoon. Anderson has found water temperatures approaching 35°C under these circumstances.[34] In the tropics, increased evaporation in the shallow intertidal areas caused by high summer temperatures, together with evapotranspiration, creates high-salinity water and can produce reverse salt wedge conditions.[17]

During exposure, the surface mud heats up, water evaporates, and the water content of the surface decreases. Anderson reports a decrease of 8% in water content.[35] This should decrease the mobility of the bed since erodability is correlated with water content. Under extreme conditions drying out and cracking of the mudflats can occur. This happens most often over neap tides on the upper flats, which remain uncovered for many tides. Additionally, increased temperature affects the strength of the electrostatic cohesive bonds, and hence, also the erodability. Experiments have shown that the erosion rate constant rises by about 50% between 20 and 30°C.[36]

Theoretically, an increase in temperature should increase the settling velocity of particles because of its effect on water density and viscosity. Owen[37] considered the effect to be mainly on viscosity, but Burt[38] states that the consequences are small compared with the effects of salinity and concentration on fall velocity. Temperature increase should also increase the thickness of the laminar sublayer, and thus decrease the ambient shear stress.[39] However, this effect should be small compared with that of biological activity.

The effect of temperature on settling velocity has been quoted as the reason for a negative correlation between suspended sediment concentration and the annual temperature variation, because the maximum concentration occurred in the winter.[40] However, this is probably related to biological activity.[41] The opposite can also occur, with a positive correlation with temperature.[42,43] Increasing temperature increases the biological production, particularly in surface films of diatoms. These increase the erosion resistance of the bed and can trap sediment in summer.[44] When the diatoms die off in colder weather the intertidal areas become more prone to erosion. Algae such as Enteromorpha are also involved, with die-off resulting in a wintertime erosion of 6 cm, balanced by a similar deposition in the summer.[41] The opposite effect seems to occur subtidally where bioturbation by worms, etc. is more important. Increased activity in summer creates higher water contents and lower critical erosion velocities. The opposite occurs in winter.[45]

Filter feeders extract sediment from suspension and deposit it on the bed in packaged form as pseudofeces. Rhoads has shown that various shellfish can filter and deposit 0.03 to 2.0 g of sediment per individual per day.[45] At this rate it must take only a few weeks to process all of the suspended sediment in an estuary. At a higher temperature more individuals will be present, which would decrease the turnover time. However, eventually there must be an effective limit when availability of food restricts further population increase.

Generally, the density of the bed decreases with increasing numbers of individual animals, creating hydraulic roughness and increasing mobility. This is offset by mucal binding. For fine sand, the combination of microbial adhesion and increased roughness at least doubles the critical erosion bed shear stress.[46] The overall effect is generally for biological stabilization of the bed.[47] However, the effects are complicated and are poorly quantified.

Nevertheless, biological processes act as major controls on sediment mobility and suspension. It would seem that temperature could exert a considerable control on the magnitude and timing of these processes. For instance, a 1°C increase in summer temperature would double the biomass of Spartina in Essex, (A. Gray, personal communication), and this would have a large effect on intertidal sediment trapping.

RAINFALL

The direct effect of rainfall in estuaries is limited to very shallow water and to the exposed intertidal areas. Any rain falling on the water is rapidly mixed by wind- and wave-induced mixing, though initially it may form a thin stratified surface layer. Experiments have shown that raindrops cause suspension from the bed in 10 cm of water,[48] but normally it is difficult to separate the effect of rainfall from that of waves. However, Anderson concluded that increased sediment concentration observed 15 cm above the bed in 90 cm of water was caused by rain.[49]

In the intertidal area the direct impact of raindrops can cause pitting, water logging of the surface, and shallow runoff. Pitting will increase the hydraulic roughness and hence the potential mobility of the mud when re-covered by water. An increase in water content of the surface mud will produce a similar effect. The runoff of heavy showers can be effective in creating small rills and gullies,[50] and carries high concentrations of sediment into the leading edge of the water surface. These processes do not seem to have been studied much, so it is difficult to scale their importance under changing climatic conditions. Nevertheless, though it may be of significance locally and temporarily, rainfall is likely to be of secondary importance overall.

WIND

Increased windiness will have the effect of enhancing the wave climate within estuaries, and will affect sea level through surge generation. Wind also affects the intensity and even the structure of estuarine circulation. Elliott has described how a down-estuary wind will enhance the surface outflow and bottom inflow in a partially mixed estuary.[51] The classical circulation was only observed for 43% of a year's recording. An up-estuary wind reversed the circulation for 21% of the time. Total inflow and outflow also accounted for a significant percentage of the time, and these events were caused by a set-down or set-up created by Ekman transport on winds blowing parallel to the coast.[52] As a consequence, a change in the prevailing wind patterns will have an effect on the estuarine circulation and flushing.

Wind-induced velocity shear in the water column and wave motion also assists vertical mixing, the vertical eddy viscosity in the surface layer being proportional to the square of the wind speed.[53]

The direct effect of the wind on intertidal mudflats is likely to be very limited. Moisture content and cohesion of muds are normally too high for direct wind-induced

Figure 5–8 The edge of a mangrove covered intertidal area affected by wave erosion created by boat traffic, West Malaysia.

movement. However, drying out occurs at high levels on the marshes over neap tides, but even there the main effect would be the release of particles deposited on the stems and leaves of the intertidal flora.

For noncohesive sandy material, wind is a significant transport process that can contribute to the form of spits at the estuary mouth. Much work has been carried out on the thresholds and rates of transport of sand, but intertidal salt can be a significant bonding agent. For 180 μm sand the threshold friction velocity varies exponentially with salt content, so that high salt content has a relatively large effect.[54]

Nevertheless, apart from its role in generating waves and in modulating the estuarine circulation, the wind effects are likely to be of minor significance.

WAVES

Within estuaries, waves are locally generated and fetch-limited. Because of this, wave motion is generally of reduced importance in the main channels, but can be of extreme importance in the shallow intertidal areas. Wave periods are short, generally only a few seconds, and though wave heights are not large, they can be depth-limited. Nevertheless, waves of a height of only 5 cm can cause erosion of intertidal mudflats.[55] Figure 5-8 shows the effect of boat wake erosion on a mangrove-covered intertidal area.

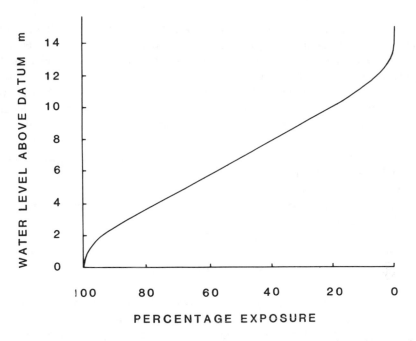

Figure 5–9 Curve showing the duration of exposure of the intertidal mudflats with water level elevation, Avonmouth, Severn Estuary. (After Shaw.[93])

The effect of waves in mixing stratification has already been mentioned.

Little work has been done specifically on wave-induced sediment movement in estuaries. However, considerable laboratory and theoretical work has been carried out to examine the processes of bed shear-stress enhancement by waves. Sand material will move at combined stresses equivalent to the Shields threshold value. For mud the wave motion creates liquifaction and movement at stresses considerably below that for a steady current. This produces a high concentration near bed suspension that has non-Newtonian viscous properties, which modifies the stresses, and can attenuate wave motion fairly rapidly.[56,57] The presence of flora has been shown to similarly attenuate waves. Wave heights can be reduced by up to 71% and wave energy by 92% by Spartina.

Wave erosion also creates a zone of high concentration at the leading edge of the tidal water level as it progresses backwards and forwards across the tidal flats. This is a likely mechanism for entraining sediment into the turbidity maximum and also for transporting sediment towards the high-water mark. Increased wave activity would increase this transport. Since high waves are often associated with surges and wave set-up, this mechanism could cause increased deposition at the highest marsh levels. The first summer storm in a Korean estuary erodes 5 to 10 cm of mud.[58] In Barataria Bay approximately 40% of all sedimentation in 1975 to 1979 was related to hurricane/storm activity, but in nonhurricane years 70 to 80% of deposition occurred during winter fronts.[59]

The extent of wave action depends on the duration of the tide exposure at various levels of the intertidal area. Figure 5-9 shows the relationship for the Severn Estuary at Avonmouth. Since the rate of change of water level is minimum at high and low water, there is a maximum duration for wave erosion.[60] Conversely, the rate of change of water level is greatest at midtide, and even though the waves cross that zone twice per tide, erosion can be considerably less. Of course, the rate of erosion will depend on the topography of the intertidal area and the width of the zone over which the wave energy is dissipated. Wave exposure is often demonstrated by grain size variation and zonal patterns on the intertidal area, with a more sandy lower foreshore.[61] The equilibrium profile is generally considered to be slightly convex, with an erosive profile being concave overall. A common feature in many estuaries is a clifflike edge to the intertidal marshes at about neap high-tide level, and this is an obvious zone of erosion in many instances.

Increasing wave incidence is therefore likely to cause significant intertidal erosion and retreat of the marsh edge.

SEA LEVEL RISE

A rise in sea level causes increased salt-wedge penetration in the estuary. For instance, a rise of 1 m can result in salt penetration by 20 km in the Delaware Estuary, thereby affecting freshwater abstraction. However, the extent of penetration may be constrained considerably by the presence of weirs at the limit of tidal influence.

Increased retention of sediment could occur in the lower courses of rivers due to the reduction in the hydraulic gradient. This influence is suggested by the study of Nichols.[62] He examined the relative sea level rise in 22 lagoons, and found that the accumulation rates in most were close to sea level rise. This indicates that the rate of submergence may control the rate at which sediment is released into the estuary from the river.

Though general sea level rise is predicted as being about 1 cm/year, sea level rise may be amplified in some estuaries, because of the changed balance between friction and funneling of tidal energy. In the Severn Estuary, modeling has shown that a 1.0-m sea level rise would give 1.1 m at Avonmouth, and 1.2 m at Sharpness, together with slight phase advances.[63]

However, more dramatic sea level changes can occur because of oceanographic features such as ENSO events. These are caused by wind changes over the western Pacific that create increased water levels which travel eastward along the equator towards America. These typically raise water levels along the west coast of the U.S. by 10 to 20 cm. However, the 1982 to 1983 event caused a 60-cm rise within 12 months on the Oregon coast.[64] Such a rise causes changes to occur without the ability of the estuary to gradually adjust and accommodate the changes. Consequently, extreme rates of erosion were experienced.

The increasing sea level also affects the response of the estuary to extremes. At a typical East Coast U.K. port a 15-cm increase in sea level doubles the probability of a sea wall being overtopped, and a 30-cm rise quadruples the risk. For lower tidal ranges the same sea level rise will produce relatively larger increases.

Radiocarbon dating indicates that the large expanse of present-day tidal marshes was found no earlier than about 4000 BP, which coincides with the reduction in sea level rise from 2.5 mm/year to 1.0 mm/year. Before that time, the rapid transgression prevented marsh growth. However, since then marsh accretion rates have been able to equal or exceed sea level rise, and marshes have grown.[65]

There are three responses of a marsh to sea level rise: the marsh can drown, remain stable, or grow.[66] If the supply of sediment is insufficient to keep up with the changing sea level, the increased frequency of tidal inundation increases salinity, and decreases the oxygen supply to the roots.[67] There is a consequent reduction in plant biomass, a loss of deposited organic material, and a reduction in the sediment-trapping capability of the marsh. Loss starts by formation of ponds or pans. Initially the retention of shallow patches of water during low tide affects the flora, which decreases the rate of accumulation. The ponds deepen and are widened by wave action, and the ponds join and eventually fragment the marsh. Stability in area implies that the marsh encroaches landward at a rate equal to that lost at the outer margins. The increase in sea level causes the tidal volume to increase faster than the cross-sectional area, resulting in higher velocities. The sediment-transport capacity thus increases, with more shoreward transport on the flood tide. With increased water depth there may be a tendency for meanders in the estuary to straighten.[68] On shoal areas the increased water depth results in increased sedimentation which leads back to an equilibrium situation, but with a time lag. As a consequence, the intertidal areas will regress as the sea level rises, but retain the same extent. However, the presence of dikes or other limiting defenses will lead to a gradually reducing intertidal area. For this reason the eastern Wadden Sea will lose 10% of intertidal area for each 0.1 cm rise in sea level.[69] Also, the North German Flakstrom has reduced its capacity by 5% between 1936 and 1982.[70]

Expansion can result from increased sediment supply. The highest rates of accumulation occur at about midtide level.[65,70] Once the marsh is above MLW, marsh grasses colonize and stabilize the sediment and provide organic material that enhances the sedimentation. Plant zonation depends upon the salinity, tidal range, and duration of flooding. Plant distribution depends upon substrate composition, soil oxygen potentials, nitrogen limitations, and interspecific competition.[71] However, most studies conclude that sediment supply limitations are the crucial factor.[59]

For the Norfolk marshes, the main source for accretion is the eroding marsh edge.[72] The lower marshes have high sedimentation rates, which relates to inundation frequency and duration. However, higher rates occur near creek margins for third- and higher order creeks.[73]

There appears to be a succession of growth stages in a marsh. The young stage has a planar front face without any creeks. Creeks and levees develop at the mature stage, and pans, cliffs, and channel meandering appear in old marshes. The sequence is from overall accumulation, through equilibrium, to overall erosion. A rise in sea level will accelerate this sequence. There are, of course, also seasonal changes with marshes exhaling suspended sediment in the summer and inhaling in the winter,[43] though other marshes have a different phasing.

Pethick has considered the growth of the Norfolk marshes to sea level, assumed to have a constant elevation.[74] He found that the rise in marsh elevation followed an exponential curve $h = a - be^{-ct}$, where h is the height of the marsh, t is time, and a, b, and c are constants with values 2.385, 1.411, and 0.014041 respectively. The

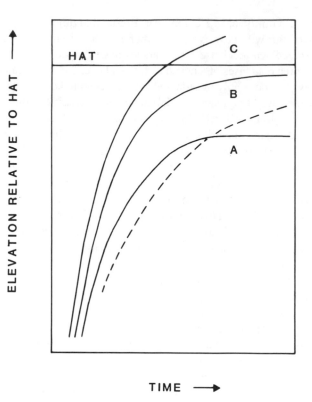

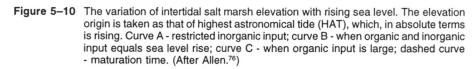

Figure 5–10 The variation of intertidal salt marsh elevation with rising sea level. The elevation
origin is taken as that of highest astronomical tide (HAT), which, in absolute terms
is rising. Curve A - restricted inorganic input; curve B - when organic and inorganic
input equals sea level rise; curve C - when organic input is large; dashed curve
- maturation time. (After Allen.[76])

constant, a, is the asymptote to which the marsh elevation rises, and the value
2.385 m corresponds to a level 0.81 m below the maximum equinoxial spring tide –
coinciding with the modal height of the tidal distribution. The time scale of the
response was 150 to 200 years. A similar function held for marshes in The Wash.[75]

Allen has taken the same concepts and applied them to the situation of rising sea
level, in a generalized form.[76,77] He assumes that the rate of change in elevation of
the marsh surface depends on sea level rise, the rates of organic and inorganic
deposition, and the rates of sediment compaction. The rate of addition of inorganic
material is assumed to be related to the elevation, falling to zero when the marsh
height reaches the height of the highest astronomical tide (HAT). Organic deposition,
on the other hand, can proceed even after the marsh has emerged above the HAT
level. Consequently, if compaction is assumed negligible, when the rate of organic
deposition is less than the sea level rise, the marsh will reach a constant elevation
with respect to increasing sea level, but at some lower level. Figure 5-10 shows such
a curve (A) plotted relative to the rising sea level elevation. Alternatively, when the

rate of organic deposition exceeds the rate of sea level increase, the marsh height can grow to outpace the increasing sea level (curve C). In this situation peat can form. When organic deposition equals sea level rise, the marsh height can eventually grow to the same elevation as sea level, and keep pace with it (curve B). The time for this to happen, the maturation time, is likely to be very long, however. For the Severn Estuary, Allen estimates maturation times of 100 to 1000 years for rates of sea level rise of 10 and 1 mm annually, respectively.[76]

The relative heights of sea level and the marsh surface are consequently sensitive to the rate of sediment input and the rate of sea level rise. An acceleration in the rate of sea level rise will lead to a relative downward movement of the marsh surface, and vice versa. Additionally, an increase in inorganic sediment supply allows the marsh to more quickly reach its mature level.

TIDAL RANGE

A change in sea level elevation can cause a change in the range of tide, particularly in high-tidal-range estuaries where the frictional dissipation of the tidal wave energy is related to the water depth. Additionally, there are long-term changes in the tides. At Newlyn in Cornwall, high-water level has increased on average by 20 cm over the past century, while low waters have increased by only 16 cm. Over this period the increase of sea level has been 18 cm. This change has been interpreted as a secular trend in tidal amplitude.

An increase in tidal range would increase the tidal velocities. This would cause increased vertical mixing, and a tendency for the estuary to become less stratified, with a decrease in the intensity of the vertical gravitational circulation.[78] However, an increase in the tidal velocities will cause sediment transport and a topographic adjustment. It has been empirically found that the mean cross-sectional area of an equilibrium estuary is linearly related to the volume between high and low water, the tidal prism.[79] Consequently, an increase in tidal range should cause a deepening of the estuary.

Also, it has been found that the cross-sectional area of estuaries varies exponentially with distance along the estuary. For instance, for the Ord River in Australia, $b_x = b_0 e^{-4.28(x/L)}$, where b is the breadth and b_0 the breadth at the mouth (x = 0).[80] Also the water depth $h_x = h_0 e^{-2.76(x/L)}$. Thus, other things being equal, an increase in tidal range will lead to an increase in breadth towards the estuary mouth, probably by the creation of wider marshes, since marsh width is correlated with tidal range.

It is feasible also that an increase in sea level could reduce the tidal range in an estuary by "de-tuning" the tidal response of the coastal sea.

Increasing tidal range increases the magnitude of the turbidity maximum.[29,81] The mass of sediment in suspension in the Severn Estuary varies dramatically with tidal range, being 7 kg m^{-3} at spring tides and 1.5 kg m^{-3} at neap tides.[63] Similar results are shown for the Ganges-Brahmaputra system with spring tide suspension concentrations being twice that at neaps.[82] One consequence of these concentration variations, when coupled with the lunar tidal residual velocity changes, is that large accumulations of fluid mud are formed at neap tide.[83] An increase in range would make fluid muds occur less often on the lunar time scale. Also, the Tay Estuary exports suspended sediment on neap tides, and imports on the spring tide.[27] Presumably, a change in tidal range would change the overall balance of export minus import.

The effect of a reduction in tidal range has been shown for the Osterschelde system, where the building of a surge barrier has decreased the range by about 12%, the tidal volume by 30%, and the currents by 30 to 80%.[84] As a consequence mud started accumulating in the channels, and on mussel beds, with sedimentation rates on the latter being about doubled. Due to the works, the ebb and flood volumes are now not in equilibrium with the cross-sectional area, and since sand does not now move within the system, the channels are being floored with mud. At present accumulation rates it is estimated that it would take 50 years to reach equilibrium. However, it is likely that as depth decreases, velocities will increase and the sediments become sandy again. The system now imports fine sediment from the sea, whereas before the tidal reduction, it exported on the order of 1×10^9 kg/year. The reduction in tidal range has excluded some marshes from inundation. The bulk density of those sediments has increased, with a consequential subsidence of 1 to 8 cm. Changes in the redox and pH profiles in the sediment and in the flora have resulted.[85]

Under natural circumstances such a drastic change in tidal range is unlikely, and the response to sea level changes is of more concern.

What seems to be an important factor in the overall response of sediment deposition in estuaries to sea level rise is the degree of tidal asymmetry within the estuary. Those estuaries with a flood dominance, a stronger flood than ebb current, tend to import sediment from the sea by the mechanism of tidal pumping. Ebb dominance leads to an export of sediment. As shown by the example of the Osterschelde, a change in tidal range or sea level may change the estuary from being a flood-dominant to an ebb-dominant system. The processes behind this have been examined in terms of tidal harmonics[86] and estuarine morphology.[87] Figure 5-11 shows that the controls are the ratios of tidal range to water depth, and intertidal volume to channel volume, though the former appears more critical. Thus, a change in water depth, without a change in tidal range, would appear to make the estuary more prone to being ebb dominant. Thus, less sediment is likely to be imported to provide the deposition required to maintain intertidal mudflats and marshes. An increase in tidal range seems only to be important in estuaries with a relatively small intertidal area and tidal amplitude.

CONCLUSIONS

Cultural activities have destroyed more coastal wetland than natural processes.[71] The U.S. has lost 0.5×10^6 acres of wetland a year from the mid-1950s to the mid-1970s, much of which was caused by waste disposal and land reclamation. Drastic changes have occurred in San Francisco Bay because of man-made changes in river flow and sediment discharge.[23,88] In the Mississippi Delta the loss of land area has grown to 100 km²/year, mainly because of the reduction in sediment input.[89] Suspended sediment input has declined 50% since the early 1970s.[90] Part of this loss of

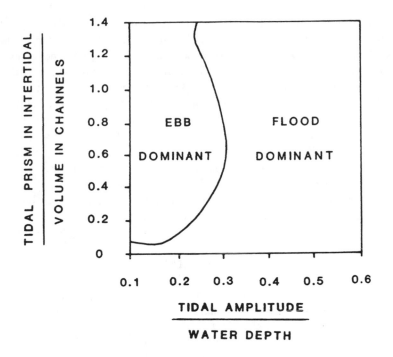

Figure 5–11 Ebb and flood dominance in estuaries, related to their volumetric and tidal characteristics. (After Friedrichs and Aubrey.[86])

sediment is due to damming upstream, but part is due to flood control measures that restrict the sediment availability to the back marsh areas.[91,92] The Mississippi, in some ways, is a good analog for future sea level rise scenarios for the larger estuarine systems, since the combination of isostatic sinking, sediment compaction, and sea level rise is creating a relative sea level rise of about 1 cm/year. Prevention of further degradation of the delta obviously requires sensitive management and the avoidance of restricting too much the natural extremes that provide the main part of the sediment input. The likely increase in rainfall and river discharge will perhaps offset the worst effects of man-induced flow reductions.

The main conclusion from this review is that the major effects of climate change appear to be manifest through changing the magnitudes of sediment inputs to the estuary, and the transport, erosion, and deposition rates within the estuary. Intertidal marshes form the zone most sensitive to change in the important variables considered, marshes are crucial in the health of estuarine systems. It is obvious in the above review that at the moment man is having a very much more drastic effect on estuaries than natural changes. Since estuaries are one of the most productive ecosystems, they are valuable environments to protect.

Estuaries are numerous when sea level is rising, but are scarce when it falls.[1] Though an estuary may be ephemeral, the biological niche that it provides through its low-salinity zone remains. Thus a new estuary will occur when conditions are right. The critical factor in determining whether the change is damaging or not is probably determined by the rate of change. Thus, the first priority in coping with the effect of climate change is to diminish man's direct effect on the estuarine environment, and allow the natural balances to be reestablished. Obviously, not all estuaries can be maintained in this way, but enough should be allowed to evolve naturally to sustain the diversity necessary for protecting the earth's natural resources and health. A "set-back" policy has been proposed for beaches where human habitation is affected by sea level rise. A similar policy is needed for estuaries. Though the cost implications would be large, this has to be set against the intrinsic and economic value of these systems. Other management scenarios include: taking no action and letting erosion take its course; artificially raising areas prone to inundation; and enclosure and protection.[94] The debate about which option needs to be taken will undoubtedly be at local level where the individual site characteristics can be properly appraised.

REFERENCES

1. Schubel, J. R. and D. J. Hirschberg. Estuarine graveyards, climatic change, and the importance of the estuarine environment, in *Estuarine Interactions*, M. Wiley, Ed. (New York: Academic Press, 1978), pp. 285–303.
2. Weisberg, R. H. A note on estuarine mean flow estimation. *J. Mar. Res.* 34: 387–394 (1976).
3. Dyer, K. R. *Estuaries: A Physical Introduction.* (Chichester: John Wiley, 1973), p. 140.
4. Hansen, D. V. and M. Rattray, Jr. New dimensions in estuary classification. *Limnol. Oceanog.* 11: 319–326(1966).
5. Prandle, D. Structure of residual flows in narrow tidal estuaries. *Estuarine Coastal Shelf Sci.* 20: 615–636 (1985).
6. Jay, D. A. and J. D. Smith. Residual circulation in and classification of shallow, stratified estuaries, in *Physical Processes in Estuaries*, J. Dronkers and W. Van Leussen, Eds. (Berlin: Springer-Verlag, 1988), pp. 21–41.
7. Bowden, K. F. and S. H. Sharaf el Din. Circulation, salinity and river discharge in the Mersey Estuary. *Geophys. J. R. Ast. Soc.* 10: 383–400 (1966).
8. West, J. R. and D. J. A. Williams. An evaluation of mixing in the Tay Estuary. *Proc. 13th Coastal Eng. Conf.* (American Society of Civil Engineers 1072), pp. 2153–2169.
9. Ketchum, B. H. Hydrographic factors involved in the dispersion of pollutants introduced into tidal waters. *J. Boston Soc. Civ. Eng.* 37: 296–314 (1950).
10. Dyer, K. R. and P. A. Taylor. A simple, segmented prism model of tidal mixing in well-mixed estuaries. *Estuarine Coastal Mar. Sci.* 1: 411–418 (1973).
11. Wood, T. A. A modification of existing simple segmented tidal prism models of mixing in estuaries. *Estuarine Coastal Mar. Sci.* 8: 339–347 (1979).
12. L'Vovich, M. I. *World Water Resources and Their Future*, R. L. Nace. Ed. (Washington, D.C.: American Geophysical Union, 1947), p. 416.
13. Halim, Y. Impact of human alterations of the hydrological cycle on ocean margins, in *Ocean Margin Processes in Global Change*, R. F. C. Mantoura, J. M. Martin, and R. Wollast, Eds. (Chichester: John Wiley, 1991), pp. 301–327.

14. Bradley, P. M., B. Kjerfve, and J. T. Morris. Rediversion salinity change in the Cooper River, South Carolina: Ecological implications. *Estuaries* 13: 373–379 (1990).

15. Lepage, S. and R. G. Ingram. Salinity intrusion in the Eastmain River Estuary following a major reduction of freshwater input. *J. Geophys. Res.* 91: 909–915 (1986).

16. Nichols, M. M. Response and recovery of an estuary following a river flood. *J. Sediment. Petrol.* 47: 1171–1186 (1977).

17. Barusseau, J. P., E. H. S. Diop, and J. L. Saas. Evidence of dynamics reversal in tropical estuaries, geomorphological and sedimentological consequences (Salun and Casamance Rivers, Senegal). *Sedimentology* 32: 543–552 (1985).

18. Zedler, J. B. and P. A. Beare. Temporal variability of salt marsh vegetation: the role of low salinity gaps and environmental stress, in *Estuarine Variability*, D. A. Wolfe, Ed. (Orlando: Academic Press, 1986), pp. 295–306.

19. Walling, D. E. and B. W. Webb. The reliability of suspended load data. *Proc. Symp. Erosion Sediment Transp. Meas.* (IAHS Publication 133, 1981), pp. 177–194.

20. Schubel, J. R. Effects of tropical storm Agnes on the suspended solids of the Northern Chesapeake Bay, in *Suspended Solids in Water*, R. J. Gibbs, Ed. (Plenum Marine Science 4, 1974), pp. 113–132.

21. Phillips, J. D. Fluvial sediment delivery to a coastal plain estuary in the Atlantic drainage of the United States. *Mar. Geol.* 98: 121–134 (1992).

22. Wood, P. A. Control of variation in suspended sediment concentration in the River Rother, West Sussex, England. *Sedimentology* 24: 437–445 (1977).

23. Krone, R. B. Sedimentation in the San Francisco Bay system, in *San Francisco Bay; The Urbanized Estuary*, T. J. Conomos, Ed. (San Francisco: AAAS, 1979), pp. 85–96.

24. Nichols, F. H., J. E. Cloern, S. N. Luoma, and D. H. Peterson. The modification of an estuary. *Science* 231: 567–573 (1986).

25. Dyer, K. R. Fine sediment particle transport in estuaries, in *Physical Processes in Estuaries*, J. Dronkers and W. Van Leussen, Eds. (Berlin: Springer-Verlag, 1988), pp. 295–310.

26. Uncles, R. J., R. C. A. Elliott, and S. A. Weston. Observed fluxes of water, and suspended sediment in a partly mixed estuary. *Estuarine Coastal Shelf Sci.* 20: 147–167 (1985).

27. Doberiener, C. and J. McManus. Turbidity maximum migration and harbour siltation in the Tay Estuary. *Can. J. Fish. Aquat. Sci.* 40 (Suppl. 1): 117–129 (1983).

28. Uncles, R. J., J. A. Stevens, and T. Y. Woodrow. Seasonal cycling of estuarine sediment and contaminant transport. *Estuaries* 11: 108–116 (1988).

29. Avoine, J., G. P. Allen, M. Nichols, J. C. Salomon, and C. Larsonneur. Suspended-sediment transport in the Seine Estuary, France: effect of man-made modifications on estuary-shelf sedimentology. *Mar. Geol.* 40: 119–137 (1981).

30. Nichols, M. M. Consequences of sediment flux: escape or entrapment? *Rapp. P-V Reun. Cons. Int. Explor. Mer.* 186: 343–351 (1986).

31. Schubel, J. R. and H. H. Carter. The estuary as a filter for fine-grained suspended sediment, in *The Estuary as a Filter*, V. S. Kennedy, Ed. (Orlando: Academic Press, 1984), pp. 81–105.

32. D'Anglejan, B. and J. Basmadjian. Change in sedimentation following river diversion in the Eastmain River (James Bay), Canada. *J. Coastal Res.* 3: 457–468 (1987).

33. Kuo, A. Y., K. Park, and M. Z. Moustafa. Spatial and temporal variabilities of hypoxia in the Rappahannock River, Virginia. *Estuaries* 14: 113–121 (1991).

34. Anderson, F. E. How sedimentation pattern may be affected by extreme water temperatures on a northeastern coastal intertidal zone. *Northeast. Geol.* 1: 122–132 (1979).

35. Anderson, F. E. The northern muddy intertidal: a seasonally changing source of suspended sediments to estuarine waters - a review. *Can. J. Fish. Aquat. Sci.* 40 (Suppl. 1): 143–199 (1983).

36. Ariathurai, R. and K. Arulanandan. Erosion rates of cohesive soils. *J. Hydraul. Div. Proc. Am. Soc. Civ Eng.* 104: 279–283 (1978).

37. Owen, M. W. *The Effect of Temperature on the Settling Velocities of an Estuary Mud.* (Hydraulics Research Report No. INT 106, (1972).

38. Burt, T. N. Field settling velocities of estuarine muds, in *Estuarine Cohesive Sediment Dynamics*, A. J. Mehta, Ed. (Berlin: Springer-Verlag, 1986), pp. 126–150.

39. Boon, J. D. Optimized measurement of discharge and suspended sediment transport in a salt marsh drainage system. *Mem. Inst. Geol. Basin D'Aquitaine.* 7: 67–73 (1974).

40. Jackson, W. H. *An Investigation into Silt Suspension in the River Humber.* (Dock and Harbor Authority XLV: 526, 1964).

41. Frostick, L. E. and I. N. McCave. Seasonal shifts of sediment within an estuary mediated by algal growth. *Estuarine Coastal Mar. Sci.* 9: 569–576 (1979).

42. Krauter, J. N. and R. L. Wetzel. Surface sediment stabilization-destabilization and suspended sediment cycles on an intertidal mudflat, in *Estuarine Variability*, D. A. Wolfe, Ed. (Orlando: Academic Press, 1986), pp. 203–223.

43. Gardner, L. R., L. Thombs, D. Edwards, and D. Nelson. Time series analyses of suspended sediment concentrations at North Inlet, South Carolina. *Estuaries* 12: 211–222 (1989).

44. Paterson, D. M. Short-term changes in the erodibility of intertidal cohesive sediments related to the migratory behaviour of epipelic diatoms. *Limnol. Oceanog.* 34: 223–234 (1989).

45. Rhoads, D. C. Organism-sediment relations on the muddy sea floor. *Oceanogr. Mar. Biol. Annu. Rev.* 12: 263–300 (1974).

46. Self, R. L., A. R. M. Nowell, and P. A. Jumars. Factors controlling critical shears for deposition and erosion of individual grains. *Mar. Geol.* 86: 181–199 (1989).

47. Montague, C. L. Influence of biota on erodibility of sediments, in *Estuarine Cohesive Sediment Dynamics*, A. J. Mehta, Ed. (Berlin: Springer-Verlag, 1986), pp. 251–269.

48. Green, T. and D. Houk. The resuspension of underwater sediment by rain. *Sedimentology* 27: 607–610 (1980).

49. Anderson, F. E. Observations of some sedimentary processes acting on a tidal flat. *Mar. Geol.* 14: 101–116 (1973).

50. Bridges, P. H. and M. R. Leeder. Sedimentary model for intertidal mudflat channels, with examples from the Solway Firth, Scotland. *Sedimentology* 23: 533–552 (1976).

51. Elliott, A. J. Observations of the meteorologically induced circulation in the Potomac Estuary. *Estuarine Coastal Mar. Sci.* 6: 285–299 (1978).

52. Elliott, A. J. and D- P. Wang. The effect of meteorological forcing on the Chesapeake Bay: the coupling between an estuarine system and its adjacent coastal waters, in *Hydrodynamics of Estuaries and Fjords*, J. C. J. Nihoul, Ed. (Amsterdam: Elsevier, 1978), pp. 127–145.

53. Kullenberg, G. Entrainment velocity in a natural stratified vertical shear flow. *Estuarine Coastal Mar. Sci.* 5: 329–338 (1977).

54. Nickling, W.G. The stabilizing role of bonding agents on the entrainment of sediment by wind. *Sedimentology* 31: 111–117 (1984).
55. Anderson, F. E. Resuspension of estuarine sediments by small amplitude waves. *J. Sediment. Petrol.* 42: 602–607 (1972).
56. Wells, J. T. and J. M. Coleman. Physical processes and fine-grained sediment dynamics, coast of Surinam, South America. *J. Sediment. Petrol.* 51: 1053–1068 (1981).
57. Wells, J. T. and G. P. Kemp. Interaction of surface waves cohesive sediments: field observations and geologic significance, in *Estuarine Cohesive Sediment Dynamics*, A. J. Mehta, Ed. (Berlin: Springer-Verlag, 1986), pp. 43–65.
58. Wells, J. T., C. E. Adams, Jr., Y- A. Park, and E. W. Frankenberg. Morphology, sedimentology and tidal channel processes on a high-tide-range mudflat, west coast of South Korea. *Mar. Geol.* 95: 111–130 (1990).
59. Childers, D. L. and J. W. Day, Jr. Marsh-water column interactions in two Louisiana estuaries. 1. Sediment dynamics. *Estuaries* 13: 393–403 (1990).
60. Carr, A. P. and J. Graf. The tidal immersion factor and shore platform development: discussion. *Trans. Inst. Br. Geog.* 7: 240–245 (1982).
61. Evans, G. Intertidal flat sediments and their environments of deposition in the Wash. *Q. J. Geol. Soc. London.* 121: 209–240 (1965).
62. Nichols, M. M. Sediment accumulation rates and relative sea-level rise in lagoons. *Mar. Geol.* 88: 201–219 (1989).
63. Severn Barrage project. General Report. Energy Paper 57. H.M. Stationery Office, London.
64. Komar, P. D. The 1982–83 El Niño and erosion on the coast of Oregon. *Shore Beach* 54: 3–12 (1986).
65. Oertel, G. F., G. T. F. Wong, and J. D. Conway. Sediment accumulation at a fringe marsh during transgression, Oyster, Virginia. *Estuaries* 12: 18–26 (1989).
66. Stevenson, J. C., L. G. Ward, and M. S. Kearney. Vertical accretion in marshes with varying rates of sea level rise, in *Estuarine Variability*, D. A. Wolfe, Ed. (Orlando: Academic Press, 1986), pp. 241–259.
67. De Laune, R. D., S. R. Pezeshki, and W. H. Patrick, Jr. Response of coastal plants to increase in submergence and salinity. *J. Coastal Res.* 3: 535–546 (1987).
68. Finkelstein, K. and C. S. Hardaway. Late Holocene sedimentation and erosion of estuarine fringing marshes, York River, Virginia. *J. Coastal Res.* 4: 447–456 (1988).
69. Peerbolte, E. B., W. D. Eysink, and P. Ruardij. Morphological and ecological effects of sea level rise: an evaluation for the Western Wadden Sea, in *Ocean Margin Processes in Global Change*, R. F. C. Mantoura, J.- M. Martin, and R. Wollast, Eds. (Chichester: John Wiley, 1991), pp. 329–347.
70. Dieckmann, R., M. Osterthun, and H.W. Partenscky. Influence of water-level elevation and tidal range on the sedimentation in a German tidal flat area. *Prog. Oceanogr.* 18: 151–166 (1987).
71. Orson, R., W. Panageotou, and S. P. Leatherman. Response of tidal salt marshes to rising sea levels along the U.S. Atlantic and Gulf Coasts. *J. Coastal Res.* 1: 29–37 (1985).
72. Reed, D. J. Sediment dynamics and deposition in a retreating coastal salt marsh. *Estuarine Coastal Shelf Sci.* 26: 67–79 (1988).
73. Stoddart, D. R., D. J. Reed, and J. R. French. Understanding salt-marsh accretion, Scolt Head island, Norfolk, England. *Estuaries* 12: 228–236 (1989).

74. Pethick, J. S. Long-term accretion rates on tidal salt marshes. *J. Sediment. Petrol.* 51: 571–577 (1981).
75. Kestner, F. J. T. The loose boundary regime of the Wash. *Geog. J.* 141: 389–414.
76. Allen, J. R. L. Constraints on measurement of sea-level movements from salt marsh accretion rates. *J. Geol. Soc. London* 147: 5–7 (1990).
77. Allen, J. R. L. Salt-marsh growth and stratification: A numerical model with special reference to the Severn Estuary, south west Britain. *Mar. Geol.* 95: 77–96 (1990).
78. Bowden, K. F. and R. M. Gilligan. Characteristic features of estuarine circulation as represented in the Mersey Estuary. *Limnol. Oceanog.* 16: 490–502 (1971).
79. O'Brien, M. P. Equilibrium flow areas of inlets on sandy coasts. *J. Wat. Harb. Div. Am. Soc. Civ. Eng.* WW1: 43–52 (1969).
80. Wright, L. D., J. M. Coleman, and B. G. Thorn. Processes of channel development in a high-tide range environment: Cambridge Gulf-Ord River Delta, Western Australia. *J. Geol.* 81: 15–41 (1973).
81. Allen, G. P. Study of the sedimentary processes in the Gironde Estuary. *Mem. Inst. Geol. Basin D'Aquitaine* No. 5, p. 314 (1973).
82. Barua, D. K. Suspended sediment movement in the estuary of the Ganges-Brahmaputra-Meyhna River system. *Mar. Geol.* 91: 243–253 (1990).
83. Kirby, R. and W. R. Parker. Distribution and behaviour of fine sediment in the Severn Estuary and Inner Bristol Channel, U.K. *Can. J. Fish. Aquat. Sci.* 40 (Suppl. 1): 83–95 (1983).
84. Ten Brinke, W. B. M., J. Dronkers, and J. P. M. Mulder. Fine sediments in the Oosterschelde tidal basin before and after partial closure. *Hydrobiologia: 282/283*: 41–56 (1994).
85. Vronken, M., O. Oenema, and J. Mulder. Effects of tide range alterations on salt marsh sediments in the Eastern Scheldt, S.W. Netherlands. *Hydrobiol. Bull.* 195: 13–20 (1990).
86. Friedrichs, G. T. and D. G. Aubrey. Non-linear tidal distortion in shallow well mixed estuaries: a synthesis. *Estuarine Coastal Shelf Sci.* 27: 521–546 (1988).
87. Dronkers, J. Tidal asymmetry and estuarine morphology. *Neth. J. Sea Res.* 20: 117–131 (1986).
88. Rosengurt, M. A., M. J. Herz, and M. Josselyn. The impact of water diversions on the river-delta-estuary-sea ecosystems of San Francisco Bay and the Sea of Azov, in *San Francisco Bay: Issues, Resources, Status and Management.* NOAA Estuary-of-the-Month seminar Series No. 6, pp. 35–61.
89. Day, J. W., Jr. and P. H. Templet. Consequences of sea level rise: implications from the Mississippi Delta. *Coastal Manage.* 17: 241–257 (1989).
90. Turner, R. E. and Y. S. Rao. Relationships between wetland fragmentation and recent hydrologic changes in a deltaic coast. *Estuaries* 13: 272–281 (1990).
91. Wells, J. T. and J. M. Coleman. Wetland loss and the subdelta life cycle. *Estuarine Coastal Shelf Sci.* 25: 111–125 (1987).
92. Salinas, L. M., R. D. De Laune, and W. H. Patrick. Changes occurring along a rapidly submerging coastal area: Louisiana, USA. *J. Coastal Res.* 2: 269–284 (1986).
93. Shaw, S. M. Class A Network Dataring gauges: 1989 data processing and analysis. Proudman Oceanographic Laboratory. Report 18 (1991), p. 134.
94. ASCE Task Committee. Effects of sea level rise on bays and estuaries. *J. Hydraul. Eng.* 118: 1–10 (1992).

CHAPTER 6

Effects of Sea Level Rise on Coastal Sedimention and Erosion

J. T. Wells

INTRODUCTION

This chapter discusses the effects of sea level rise on coastal processes and details the anticipated sedimentary responses. The goal is to provide a basis for prediction. The chapter is broadly divided into two sections. The first contains a discussion of how the fundamental driving forces and transport mechanisms are affected by rising water levels; the second is a discussion of anticipated responses in specific coastal environments that together make up the majority of the world's shorelines. Within each section, projections for the future are summarized and some of the man-induced effects are examined.

FLOODING VS. EROSION

Changes in sea level can be expected to have a profound effect on coastal sedimentary environments.[1] As sea level rises, shorelines throughout the world will respond by flooding or eroding. Each response is a type of marine transgression that leads to rapid recession or migration of the shoreline. Whereas flooding is the inundation of low-lying coastal land which is unable to build upward or outward at a rate sufficient to keep pace with sea level rise (Figure 6-1), erosion is the physical removal of sedimentary materials which form the shoreline (Figure 6-2). Erosion accounts for most of the net shoreline recession on beaches and barrier islands, such as the east coast of the U.S. and the west coast of Africa;[2-4] flooding accounts for most of the loss in wetlands and subsiding deltas such as the Mississippi, Ganges, and Niger.[5-7]

0-87371-301-X/95/$0.00+$.50

Figure 6–1 Wetland loss in the Mississippi Delta resulting from inundation. High subsidence rates and loss of sediment input due to upstream engineering structures are responsible for much of the land loss.

Figure 6–2 Beach erosion in North Carolina is a classic example of transgression resulting from the physical removal of sedimentary materials.

Specific responses will depend on many obvious physiographic considerations, such as topographic relief, erodability of the beaches, and degree of sheltering from wind, wave, and tide energy. Other less obvious factors may include rate of subsidence from fluid withdrawal and modifications from engineering structures. In terms of flooding, the most important variable is slope. Steep slopes prevent large horizontal shoreline displacements for a given sea level rise, while gentle coastal plain surfaces are more susceptible to flooding. In terms of erosion, the single most important variable is the character of shoreline material. Rocky shorelines obviously are the most resistant to erosion, followed generally by gravels and sands, then silts and clays. However, consolidated muds or muds that are densely bound by vegetation, may actually withstand erosion better than coarser sands.

Although, to a first approximation, shorelines can be divided into those that will flood and those that will erode, the processes of flooding and erosion are closely related. For example, some barrier island shorelines will initially erode, perhaps in a stepwise fashion, then flood or "drown in place" during periods of low sediment supply.[8,9] Other barriers may erode on both the seaward and landward sides until efficient overwash can take place, allowing the islands to begin migrating intact.[10] Still other shorelines may initially be submerged, then begin eroding when embayments and shallow bodies of open water become large enough to accommodate storm waves, as appears to have happened during the widespread internal fragmentation of some wetlands in the Gulf of Mexico.[6]

Flooding and erosion are responses that are also related temporally. Sea level, which has undergone huge climatically induced excursions, is clearly responsible for large-scale marine transgressions.[11-14] Retreat of the shoreface for great distances (10 to 100 km) has been driven by the potential for flooding over geologic time scales. Without sea level rise, these transgressions would not occur. Yet, it is the seasonal storm waves that produce the widespread erosion and redistribution of sediment. Storms, fortnightly tides, and seasonal wind patterns work on the shoreline at higher frequencies and are the agents of short-term sea level changes. Regardless of greenhouse-induced climatic changes, nearly all of the world's shorelines have been, and will continue to be, subject to episodic "crescendo" events that produce far greater short-term hazards than secular sea level rise.[15]

PROBLEMS IN PREDICTION

One of the chief problems in predicting shoreline response to rising sea level is our inability to produce a universally applicable model. Future rates of relative sea level rise will vary from coast to coast. This has led to at least four approaches[5] to predicting shoreline response: (1) use of historical data,[16-18] (2) application of the principle of conservation of sediment volume,[19-21] (3) application of Bruun Rule equilibrium profiles principles,[10,22-24] and (4) modeling profiles on the basis of offshore forcing.[5,25,26] Unfortunately, all but the first of these approaches have been derived for sandy shorelines and can not readily be applied to other common types of coastal environments, such as deltas, wetlands, reefs, and tidal flats.

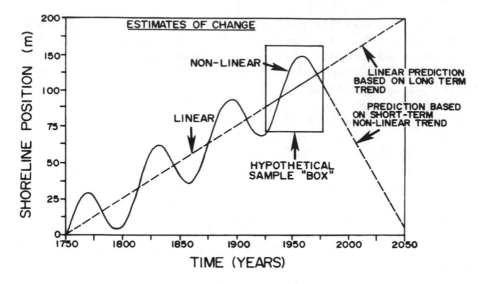

Figure 6–3 Example of a hypothetical change in shoreline position that is cyclic over a period of hundreds of years. Estimates for predicting the future trend from the 50-year sample box illustrates the difference in linear and nonlinear models: the quadratic fit is a better short-term model but the linear fit is a better long-term predictor. (Modified from Dolan, et al.[27])

Application of different approaches leads to different results on the same section of shoreline. For example, Pilkey and Davis[10] found that predicted shoreline recession from extrapolated rates on barrier islands produced values that were unrelated to results of the commonly used Bruun model. They pointed out that the possibility of an accurate model is highly unlikely because the interaction of coastal processes is not fully understood. Similarly, Dolan et al.[27] found that fitting a linear model to shoreline response would only approximate the average rate of change from shorelines with nonlinear response, but that linear fits may be more representative of trends over long time periods (Figure 6-3). They noted that trends in shoreline movement, from which predictions might be made, can be unidirectional and constant, or cyclic with accelerations and decelerations. Figure 6-3 illustrates this concept and shows the difficulty of predicting recession from past performance by using different approaches.

Another problem in prediction is that sea level rises gradually, even using the worst-case scenario of a 1- to 2-cm rise per year. Because of the slow, gradual rise, there is a lag time in response. We are not at all certain as to what the short-term response is to slow rates of sea level rise since it appears to differ from that produced by very rapid rates.[5,28] Erosion is a time-averaged phenomenon that results from storm generated site-specific events over the long term,[29] and many of the shoreline changes are in fact cyclic as shown in Figure 6-3. Separating the effects of this weak long-term sea level signal from the noisy short-term storm record is exceedingly difficult.

Finally, there is the problem of accuracy and precision. Estimates of past shoreline change can be used to predict future changes, but reliability depends on the particular method utilized, the time between measurements, the total length of the measurements, and the magnitude of shoreline position errors.[29-32] The overwhelming need for highly accurate shoreline maps and prediction models has unfortunately shifted emphasis away from processes that actually cause the shoreline movement.[18] To be effective, predictions from observed shoreline changes must always be used in conjunction with basic principles of water and sediment movement. In the following section, the effects of sea level rise on several of these basic driving forces are reviewed.

THE FUNDAMENTAL PROCESSES AND RESPONSES

WAVES AND TIDES

Despite the lack of a universally applicable recession model, knowledge of the chief driving forces is essential because these processes respond to increases in water level through reduced bottom friction. With substantial increases in sea level over long periods of time, deepening of water on the continental shelf could result in less wave damping and more energy at the shoreline. Presentation of two hypothetical cases, one for pre-existing waves and the other for locally generated waves, indicates that a 1-m rise in sea level on a shelf 10 km wide and 10 m deep would increase wave height by 3 and 7.5%, respectively.[5] Wave energy, which increases as the square of wave height, could have a substantial engineering impact in the second case.

The most significant impacts, however, are likely to occur in shallow bays, estuaries, or lagoons where an increase in sea level of say, 1 m, is a greater percentage of the total water depth. Many such sheltered coastal environments are dominated by unconsolidated fine-grained sediments and are therefore highly susceptible to erosion. As a generalization, waves break when their height reaches about 0.75 the depth of water in which they are traveling. The implication then, is that a bay created by a 1-m rise in sea level could accommodate 0.75-m-high waves in a region that was previously not flooded. Wetlands that are already in a delicate ecological state of balance[33-35] may begin their rapid decline once waters are deep enough for local wind waves to regularly erode the substrate of interior shoreline margins.

Tides also respond to bottom friction and, because of their long periods, are actually a type of shallow water wave. Tide height and tidal currents are controlled largely by basin geometry. It is therefore unlikely that a sea level rise of the magnitude anticipated on a century time scale would create any effect on the open coast. However, a rise in sea level will increase the volume of water transported into bays, lagoons, and backbarrier environments by tidal currents. This increase in the so-called tidal prism, while small in comparison to the total tidal prism, may affect inlet size and stability or promote creation of new inlets.

Within shallow estuaries and bays, an increase in sea level will generally increase the tide range unless sedimentation has been able to keep pace with the rise. Mehta et al.[25] have shown that the opposite effect is possible in special situations where tidal resonance, such as in the Bay of Fundy, is important. In most bays, however, there is a superelevation of water, caused by factors such as inlet and bay geometry, river and local runoff, winds, and wave setup effects. These physical processes might also be influenced by sea level rise[36] so as to effect tide range. In at least one instance, the German Bight, examination of long-term sea level trends has offered corroboration of a change in superelevation that was reflected by rising tide ranges.[37]

STORM SURGE

Storm surge is an increase in water level from a combination of wind stress and reduction in barometric pressure during storms. Larger surges occur in broader shelf areas and when funneled by converging shorelines within estuaries.[19] Most damage to property is produced by storm surge, either from tropical (e.g., hurricanes originating in the tropics) or extratropical (e.g., cold front or warm front) storms, because of the potential for causing abnormal water levels at the shoreline. These water levels will not only flood low-lying coastal areas, but will also provide the temporary water elevations necessary for waves to attack farther inland.

Surge from a storm of given size, intensity, track, and speed of forward motion will be increased if there are larger expanses of shallow water resulting from sea level rise. On the other hand, if the shoreline is fixed by engineering structures, then an increase in offshore water depth will tend to reduce storm surge effects because of the inverse relationship to water depth. Although a hypothetical case for a 1-m sea level rise shows that the return period for a 4-m surge could be reduced from 100 to 50 years,[5] the general inability to verify storm surge models makes accurate prediction difficult. Perhaps more significant than the effects of sea level rise on storm surge would be the potential effects of climate change on storm and hurricane frequency and intensity.[38]

SEDIMENT TRANSPORT: LITTORAL PROCESSES

Sea level rise will be a significant factor in sediment transport if it allows larger waves to reach the shoreline and if it significantly increases the elevation of wave attack on the beach. Three factors would lead to greater transport rates. First, larger waves would generate stronger longshore currents. Speed of the longshore current, which is proportional to the maximum orbital velocities of the waves, would increase with increases in wave height.[39,40]

Second, larger waves will be able to resuspend greater quantities of sediment as a result of the higher orbital velocities at the bottom. Since waves can resuspend sediment of a given grain size at much lower equivalent speeds than a current that is steady, the interaction between waves and currents is also important. This is because once waves have lifted the sediment from the bottom, it can then be transported by currents which alone would be unable to resuspend the sediment. Coupling between stronger longshore currents and additional turbulence from waves thus provides for greatly enhanced sediment transport.[41,42]

Third, during the more rapid periods of sea level rise, sediment may not be transported shoreward at sufficient rates to maintain a stable shoreline.[43] Net transport

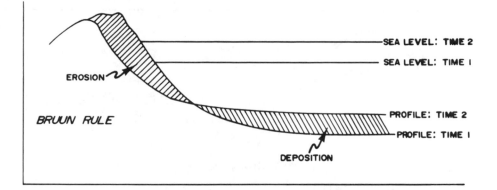

Figure 6–4 The Bruun Rule is a simple two-dimensional model that postulates, under rising sea level, erosion of sediment from the beach and shoreface and deposition on an offshore ramp. The increase in elevation of the ramp equals the amount of sea level rise.

of sediment may actually end up directed offshore since the beach profile will respond to greater turbulence in the surf zone by transferring sediment seaward to create a more gentle nearshore slope. Beach sediments are part of a sand-sharing system in which the upper beach face and even the dunes are removed during storms and temporarily stored offshore in a system of ridges and runnels.[44] This change in shoreline position should not be confused with net long-term changes since most of the sediment moves back onshore during the days to weeks following the storm. However, as sea level rises, the onshore-offshore sand-sharing system may loose its efficiency, resulting in net offshore transport of sand because of one or more of the following: an increase in wave steepness resulting from a change in breaker type; wave energy dissipation over a smaller (steeper) section of beach; and, impingement of the shoreline into finer sediments that are unstable on the beach face.

THE BRUUN RULE

Perhaps no method for predicting shoreline recession has received more attention than the Bruun Rule.[22] Simply stated, the Bruun Rule formulates the relationship between rising sea level and the rate of shoreline erosion; it shows that, for a given sea level rise, gentle beaches will erode faster than steep beaches.[22,45] The rule assumes that sediment removed during shoreline retreat is transferred offshore so that the inner continental shelf gains sediment elevation at a rate equal to the rise in sea level (Figure 6-4). An equilibrium (or average) beach profile thus maintains its original shape and moves landward and upward at a rate about 100 times the rate of sea level rise.

The Bruun Rule assumes that no sand is lost to longshore transport and that an offshore "closure depth" exists beyond which there is no sediment exchange. The rule does not allow for shoreward transport of sediment as overwash[46] nor for situations where the coastal plain slope is too gentle for sufficient sand to be available as a source for supplying the offshore. The implication of not having contributions

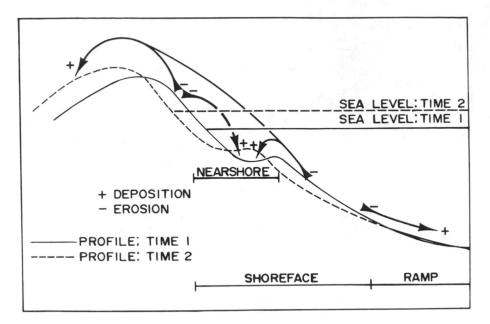

Figure 6–5 Diagram showing in two dimensions the directions of sediment transport on a beach. Sands move both onshore and offshore in response to rising sea level. Longshore transport is not shown. (Modified from Dubois.[57])

from the shelf (closure depth requirement) is that erosion is the only possible shoreline response to a sea level rise. However, during periods of low to moderate sea level rise, there appears to have been a landward rather than seaward transport of sediments (Figure 6-5), bringing into question the concept of continuous offshore transport.[43] There is little question that without the supply of shoreface sediments, transgressive barrier islands would not have been able to maintain subaerial integrity.[47,48]

Despite geometric shortcomings and the fact that the model does not incorporate evolutionary concepts into beach and barrier island processes,[10] more than two decades of field and laboratory studies have shown many cases where the Bruun Rule appears to be valid. These include wave-tank and follow-up field experiments on the eastern coast of the U.S.,[49,50] studies on the Great Lakes[23,51-54] and in low energy lakes in New Zealand,[55] and in sections of the Chesapeake Bay.[56] An international working group, brought together to assess the status of models and analytical techniques for erosion response to sea level rise, confirmed the basic patterns of the Bruun Rule but recommended that it be used only for order-of-magnitude estimates.

A recent re-evaluation of the Bruun Rule by Dubois[57] concluded that it is applicable to the beach and nearshore, but that the offshore profile, rather than being elevated in proportion to sea level rise, is simply abandoned by wave action.

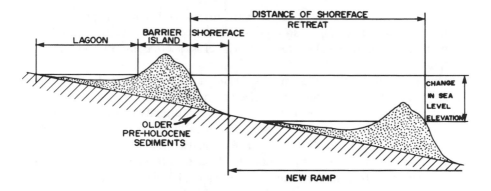

Figure 6–6 Intact migration of a barrier island across the inner continental shelf. Steeper slopes result in shorter retreat distances for a given rise in sea level. Such idealized movement maintains barrier island shape, volume, and geomorphic features such as the lagoon. (Modified from Everts.[59])

EFFECTS ON SELECTED COASTAL ENVIRONMENTS

BEACHES AND BARRIER ISLANDS

Approximately 90% of U.S. beaches, and perhaps 70% worldwide, are eroding.[4,58] Eustatic sea level rise is one cause, but other natural factors such as antecedent topography, sediment supply, and variation in waves, currents, tide, and wind are also important.

Beaches occur wherever there is sufficient supply of sediment for accumulation at the shoreline. Most are composed of sand, gravel, or abraded shell and are broadly classified as either mainland or barrier beaches. In addition to the long and continuous "strandplain", mainland beaches can take several other forms: pocket beaches between headlands (coast of Oregon); reworked deltaic distributary-mouth bar sands (South Pass, Mississippi Delta); and, chenier ridge bundles derived from updrift fluvial sources (coast of the Guianas, northeastern South America). Barrier beaches, which occur either as barrier islands or spits, differ in that they are backed by lagoons, sounds, or estuaries. Although mainland and barrier beaches are present in a wide variety of oceanographic and geologic settings, and have displayed complex and varied responses to past sea level changes, there are four fundamental principles that will likely govern their future responses.

First, a sea level rise will force sediment redistribution at the shoreface. Thus, an increase in rate of sea level rise will almost certainly cause faster erosion rates, assuming that other factors remain constant. This is not to say that most sandy shorelines are now eroding simply because of sea level changes, as Bird[2] has so clearly pointed out. Rather, the fundamental principle here is that a future increase in rate of sea level rise can be expected to drive a marine transgression, which we will recognize and identify as erosion. However, to many geologists, erosion is actually shoreface retreat of mainland beaches or, in the case of coastal barriers, just a component of natural island migration. Human development close to the shoreline gives us cause to interpret migration as erosion.

Second, slope of the migration surface is critical to the rate of recession (Figure 6-6). If shorelines are retreating up a slope, then the geometry of the slope will be a factor in the rate and distance of movement.[10] Just as the Bruun Rule predicts faster erosion rates on gentle slopes, so do other more recent techniques for predicting shoreline recession, which allow for overwash, cross shore transport, and longshore transport.[21,24,59]

Third, sediment availability may override the first-order effects of sea level change. If sediments are of the correct size and available, then beach accretion can prevail over modest rates of sea level rise.[60] This is especially true when large flood tidal deltas serve briefly as reservoirs of sediment during initial stages of island migration. However, the single largest source of sand to the coast has been rivers, and most river-borne sediment is now trapped by dams or is retained in estuaries. Other sources, such as cliff erosion, continue to supply some beaches (see next section), but much of the beach sediment comes from longshore or possibly offshore sources. Since excess sand is uncommon, delicate balances often exist and coastal accretion will be important only locally.

Fourth, whereas many barrier islands will respond to increasing rates of sea level rise by migrating, others will become thinner through simultaneous ocean and sound-side erosion, or be forced to drown in place. Prevailing thought has been that migration is the dominant process; inability to migrate by transferring sediment from the seaward to landward side of an island is usually taken as evidence that a barrier feature will eventually drown in place. However, recent research shows that on 100-year time scales, barrier islands may experience erosion on both sides, thinning down to some critical width that then allows overwash to become an efficient process.[29] Once a critical width is reached (300 to 500 m), rate of shoreline movement will increase sharply as all available sand begins moving across the island. Rapid onset of overwash following the thinning process, together with changes in sediment supply, migration slope, and frequency of storms, will lead to episodic rather than smooth and continuous response of mainland and barrier beaches (Figure 6-7).

At mid to high latitudes, beaches and barrier islands are often composed of gravel derived from Pleistocene glaciation. The most notable examples are in Great Britian, Ireland, Nova Scotia, British Columbia, and parts of Alaska. Like their sandy counterparts, gravel barriers appear to respond to rising sea level by migrating landward; however, relatively little is known about migration processes since much of the evidence is in areas of very slow (Ireland and England) sea level rise.[61] Detailed studies of gravel barriers in Nova Scotia, where relative sea level rise has been considerably faster (2 to 3 mm/year), have shown that migration is by overwash and barrier rollover processes whereby beachface sediment is passed over the barrier crest by storm waves.[61,62] Rates appear to be directly proportional to short-term rates of smoothed sea level change, and the response of gravel barriers may be more sensitive to these short term changes than previously thought. While it is likely that gravel barriers will also transgress more rapidly at higher rates of sea level rise, local variations in sediment supply and the underlying topography will compartmentalize the process and complicate simple cause-and-effect relationships.[63]

Figure 6–7 Removal of overwash sand following a northeaster storm on the east coast of the U.S. Overwash is a natural part of barrier island transgression.

SEA CLIFFS

Sea cliffs, which comprise a substantial portion of the world's shoreline,[2,64] occur in a wide variety of forms: glacial till on the eastern coast of the U.S. and in the Arctic tundra; sandstone in parts of Germany, Japan, and the U.S.S.R.; limestone near Perth, Australia; chalk along the south coast of Britian; and volcanic ash on Krakatoa Island. Average erosion rates vary from 0.001 to 0.01 m/year for limestone cliffs to 10 m/year for unconsolidated volcanics,[65] with a worldwide rate reported to be about 0.05 m/year.[64] Unlike beaches and low barrier islands, sea cliffs are not flooded during storms, and processes of overwash will not dominate shoreline recession during periods of sea level rise. Rather, shoreline recession will be controlled by lithology, geologic structure, precipitation, ground water level, and variations in average wave conditions.

Even under constant levels of wave energy, erosion of sea cliffs takes place intermittently. In a summary of sea cliff processes, Sunamura[65] has shown that there is a critical wave height at the base of a cliff for initiating erosion and that the large,

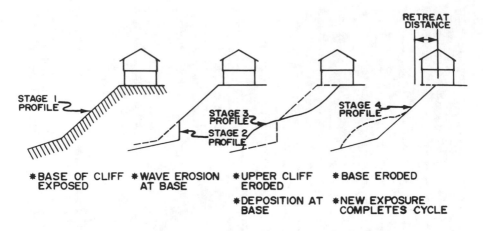

BASE OF CLIFF **WAVE EROSION** **UPPER CLIFF** **BASE ERODED**
    EXPOSED   AT BASE   ERODED

                                      DEPOSITION AT **NEW EXPOSURE
                                          BASE   COMPLETES CYCLE

Figure 6–8 Idealized stages of cliff retreat showing the alternation of erosion and deposition at the base. (Modified from Everts.[68])

less common waves are responsible for the recession. As waves undercut sediments at the foot of a cliff, the increase in slope instability leads to mass movement processes, which in turn make the profile stable by supplying debris to the foot of the cliff. Further recession is then halted until waves remove these sediments and again create an oversteepened profile. Thus, during recession, sea cliffs alternate between steep and gentle profiles, fast and slow recession rates (Figure 6-8). Since more time is needed for waves to remove large amounts of cliff debris, it is usually assumed that, under given wave conditions, higher cliffs have smaller erosion rates than lower cliffs. However, this concept has been found to be true only over very short time intervals.[66,67]

The most likely effects of an increasing rate of sea level rise are (1) generation of larger waves due to increases in water depth over the erosional surface of the shore platform, (2) a narrowing of the beaches that front sea cliffs due to the larger waves and elevated water levels, and (3) as a result, an increase in the percentage of sand supplied to the littoral sand budget by mass wasting of sea cliffs. Together these will lead to an increase in the rate of cliff erosion, and therefore in average rate of shoreline recession, since additional sediment contributions from cliff erosion will achieve a balance with the reduced beach widths.[68]

As sea level rises, sea cliffs and the beaches that front them will remain as a coupled system; the same beaches that protect sea cliffs will continue to be nourished by erosion of the cliffs. Since waves cannot attack the base of a cliff until the beach is temporarily removed during storms, this sediment supply to the beach, especially in the absence of riverine sediment sources, will be essential. However, it is unlikely that long-term prediction of recession rates will be an overriding issue in a period of rising seas. The intermittency of the process of cliff erosion, i.e., very high short-term rates but enormous variation in these rates, will overshadow processes that operate on time scales of longer than a decade. Highly developed areas will probably experience disastrous recession occurring on time scales far shorter than those required for significant increases in wave height and water level. Unfortunately, attempts to halt the process of cliff erosion by engineering structures will lead to loss of sediment supplied by the cliffs at a time when it is critically needed, further accelerating any sediment deficit.

DELTAS

Deltas are perhaps the most dynamic of all coastal landforms. On the one hand, they are the only major coastal environments which receive large, continuous influxes of sediment necessary for building new land. On the other hand, deltas are highly susceptible to the effects of sea level rise because of their low relief, erodible substrate, and high subsidence rates. Many deltas, such as the Mississippi, Ganges-Brahmaputra, and Niger have local subsidence rates that are at least 10 times the rate of eustatic sea level rise. The balance between subsidence, sea level rise, and sedimentation processes is of great economic and societal concern because agricultural activities and large population centers, especially in the tropics, are concentrated in major river basins and their deltas.

The 30 to 35 major deltas, as well as the hundreds of smaller deltas of the world, display a wide range of configurations that will affect their responses to sea level rise. In many respects deltas are composite features made up variously of beaches, spits, dune fields, tidal flats, wetlands, and active and abandoned distributaries.[69] Whereas some deltas experience low wave energy and negligible tides, others are exposed to continuous and severe wave forces or to tide ranges that may exceed 5 m. Many deltas, such as the Mississippi, are dominated by silt- and clay-sized particles, whereas others, such as the Burdekin in Australia, are composed almost exclusively of sand and gravel. The common attribute shared by each of these deltas, regardless of environmental setting, has been the ability to accumulate fluvial sediments more rapidly than they could be removed by marine processes.

However, this is no longer true in most deltas. A number of factors appear to have tipped the balance from progradation to that of recession. Even at present modest rates of worldwide sea level rise, many large marine deltas (e.g., the Mississippi, Indus, Ganges-Brahmaputra, Niger, and Nile) have entered a transgressive phase because of factors such as decreasing sediment loads, increasing subsidence, natural channel switching and lobe abandonment, and leveeing and other human activities.

It is rare that a delta progrades seaward without some lateral shift or without changing its deposition site. As river channels extend seaward they become inefficient and loose their gradient advantage, subsequently switching to a new location for dispersing deltaic sediments. The Mississippi Delta has switched depositional sites at least seven times in the past 5000 years[70] and the Hwang Ho Delta in China episodically shifted depositional sites in the mid-1800s from the Yellow Sea to the Gulf of Po Hai, some 500 km to the north. Following each shift, there is a period of coastal erosion as sediments are diverted to a new site (Figure 6-9). It may take decades to centuries before the amount of land being added to a new delta lobe offsets the amount of land being lost from an abandoned delta lobe. Since the Hwang Ho shifted its course in 1855, approximately 17 km of coastal erosion has occurred at the old depositional site.[71]

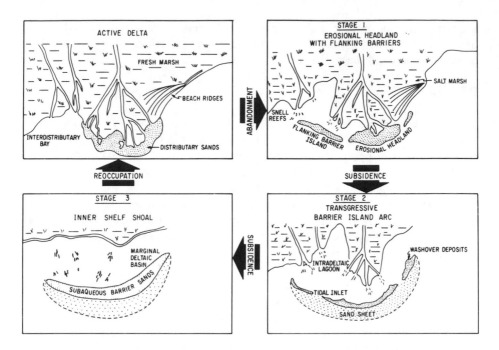

Figure 6–9   Evolutionary stages of a delta cycle. Accelerated sea level rise during a period of delta abandonment will have devastating effects. Such is the case in the Mississippi Delta. (Modified from Coleman;[75] originally from Penland and Boyd, 1981.)

The combination of increased subsidence because of fluid withdrawal and decreased sediment loads through engineering modifications, assures that virtually any acceleration in rate of sea level rise is likely to have devastating effects. The fact that many temperate and tropical deltas are covered by a fragile living surface of salt marsh or mangrove swamp (see next section) accentuates the problem. Subsidence rates in these deltas are already many times greater than rates of eustatic sea level rise, creating a situation where a delicate balance can only just be maintained; once the balance is lost, deterioration occurs rapidly, perhaps geometrically, as the living surface dies from anoxia or saltwater intrusion. One of the most striking examples is in the Mississippi Delta where predictions suggest that, even without an increase in rate of eustatic sea level rise, the delta will be entirely under water by the year 2020 (Figure 6-10).[6]

Although inundation in subsiding deltas is a worldwide problem,[7] there are numerous deltas that are on relatively stable geologic platforms (e.g., the Senegal in Africa, Ord in Australia, and Yangtze in China). These deltas will probably transgress through sediment removal at the seaward margins. Sandy deltas in extremely high-wave-energy environments, such as the Indus, will experience reworking of delta front sands in such a way as to create a transgressing beach environment.[72] In arid climates, loss of any already-sparse vegetation through saltwater penetration is likely to increase wind-blown sands and form a landscape dominated by dune

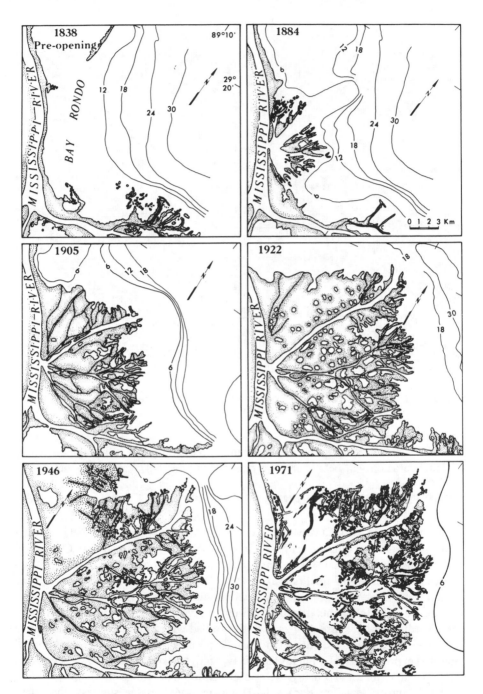

Figure 6–10 Growth and deterioration of a subdelta within the Mississippi Delta. Loss of wetlands, which form a thin living surface, are tied to the dynamics of the delta substrate.

Figure 6–11 Unvegetated tidal flat along the western coast of Korea where spring tide range exceeds 9 m. Sedimentation rates appear sufficient to keep up with present rates of sea level rise.

fields.[73-76] In a small but fortunate way, some of the sediment lost through processes of deltaic erosion will be a source for replenishing downdrift environments.

TIDAL FLATS AND WETLANDS

Tidal flats and associated wetlands occur in estuaries (Chesapeake Bay), coastal embayments (Bay of Fundy), behind barrier islands (The Netherlands and Germany), and on open coasts where wave energy is moderate to low (northeastern South America). By definition, tidal flats are low-sloping Holocene accumulations of sand or mud that grade laterally into the subaqueous environment (Figure 6-11). Most tidal flats derive their sediments directly from major rivers or indirectly from reworked deltaic sediments. Because they are best developed where tide range is greater than 2 m, the characteristic feature of tidal-flat shorelines is an extensive section of intertidal exposure (usually muddy) that is typically backed by salt marsh vegetation in temperate climates and mangrove swamps in tropical climates.

Evolution of tidal flats during the Holocene transgression suggests that these environments are able to achieve sustained vertical growth when sea level is rising at rates of at least 1 to 2 mm/year. Seaward progradation of intertidal flats depends mainly on a moderate to high sediment yield,[77] but most of the genetically related marshes appear not to have been formed until after the mid-Holocene rapid sea level rise.[8] During the past 5000 years, tidal flats in southwestern Louisiana and in northeastern South America have grown seaward as prograding mud wedges for distances of 25 and 100 km, respectively.[78,79] In both The Netherlands and in northwestern France tidal flats have grown laterally at rates of 1 to 5 m/year over the last few centuries. Even along the western coast of Korea, where local sediment sources are relatively small, tidal flats have grown vertically at very high rates: 1.5 mm to greater than 20 mm/year over the past 100 years. Present lateral progradation and vertical accumulation rates give reason to believe that tidal flats will continue to grow under modest increases in rate of sea level rise.

In contrast, considerable controversy exists over whether salt marshes will be able to survive similar increases.[34,80,81] Although marshes are usually considered sites where vertical accumulation easily keeps pace with sea level rise,[82-85] there is a net loss in many parts of the world.[34,86] This at first may seem surprising in that marsh vegetation absorbs energy and alters patterns of sedimentation by trapping particles, with roots helping secure the substrate beneath the sediment-water interface.[87]

Stevenson et al.[34] found rates of sediment accumulation at 15 marsh sites in the U.S. to vary from 1.4 to 14 mm/year. When compared to rates of apparent sea level rise, 75% of the sites had a positive accretionary balance. Much of the variabiltiy in local sea level rise, including the effects of subsidence, accounts for the variability in vertical accretion rates. However, many marshes were in a state of delicate balance and were considered on the borderline of submergence. French[88] noted that back barrier marshes in Norfolk, England would have increased sedimentation rates as inundation frequency increased, but that at rates of sea level rise exceeding 6 mm/year the marsh would fail to keep pace and drown in place in an ecological sense.

Reasons behind present-day marsh loss can be broadly grouped as (1) loss of substrate upon which marshes are built, as in the case of subsiding deltas, (2) landward barrier island migration, which forces the marshes behind them to be displaced and eventually buried or eroded, (3) loss of sediment input through up-stream engineering works such as dams and levees, (4) direct removal by man from dredge and fill activities and construction of pipeline and navigation canals, and (5) construction of bulkheads and revetments at the coast, preventing landward translation as sea level rises. Erosion takes place by one of several methods (Figure 6-12). There may be erosion of the marsh edge by waves, forming a steep cliff, or enlargement of tidal creeks by bank collapse. However, the resistance of marsh substrate to these types of erosion suggests that submergence will be the leading cause of marsh loss. The most common type of submergence results from formation of interior ponds as deterioration of vegetation occurs within the marsh because of anoxia.

With an acceleration in the rate of sea level rise, several generalizations can be made: (1) marshes that continue to accumulate will be those in sheltered locations, (2) high-tide-range marshes may be less vulnerable to erosion or flooding because of greater sediment transport potential, (3) marshes with high sediment input from outside sources will be best able to survive, (4) subsidence, which may explain 50 to 90% of the relative sea level change, will play a large role in the ability of a marsh to survive, and (5) as in other coastal environments, the patterns will be highly site specific.

Figure 6–12 Formation of a crenulated scarp due to erosion of marsh in the northern Gulf of Mexico. Stiff marsh substrate is considerably more resistant to erosion than coarser unvegetated sediments.

Although robust in many regards, extensive mangrove shorelines are also limited to calm water, gently sloping intertidal surfaces, and relatively stable sea level (Figure 6-13). Stratigraphic records, similar to those for salt marsh sediments, indicate that large mangrove swamps did not exist in the earlier Holocene, but became established as sea level stabilized. Mangrove swamps that grow in areas that receive large sediment supplies can prograde rapidly (tens to hundreds of meters per year);[89] however, the seaward movement of mangrove margins is generally considered a response to sedimentation and not a factor that controls landform evolution as previously thought.[90]

A recent review by Ellison and Stoddart[91] found that, in the absence of terrigenous sediment input, mangrove ecosystems could keep pace with rates of sea level rise of 8 to 9 mm/year, but would be under severe stress and could not survive at rates exceeding 12 mm/year. Given the habitat diversity, especially with regard to sediment budgets, a scenario of ecosystem collapse during the next 100 to 200 years is realistic only in the case of low-carbonate shorelines where sediment input is very low. Other factors, such as increased organic accumulation, increased tidal flushing, and increased wave energy could have both positive and negative effects on survivability. As in the case of salt marshes, mangroves will ultimately respond to large-scale geomorphology and sedimentation patterns (Figure 6-14).

REEFS AND ATOLLS

Like mangroves and salt marshes, coral reefs are ecosystems that occur at or near mean sea level. However, reefs differ in that they are dependent on the ability to produce their own sedimentary material that is required for vertical growth; the sediment budget depends entirely on biogenic production. Since reef growth is intimately tied to changes

Figure 6–13 Growth of mangroves in northeastern South America (Suriname) on prograding muddy substrate. Three distinct levels of growth can be identified and related to pulses of sediment from the Amazon River.

in water level, topographic highs often become preferential sites for growth. In addition to water level, growth of corals and the reefs they form depend on tectonic activity, wave energy, substrate type, nutrients, turbidity level, and water temperature and chemistry. Geologic evidence shows that, despite their rapid growth, most coral islands and intertidal reef flats are younger than 5000 years.[92]

The growth capacity of individual reef-building organisms sets absolute limits on reef growth. Massive corals have typical growth rates of 10 mm/year and maximum rates of about 20 mm/year.[93] The geologic record shows that during the period of maximum sea level rise, maximum growth rates were of the same order (14 mm/year).[94] Under conditions of sea level rise, reefs will accelerate their growth. According to Lidz and Shinn,[95] coral communities protected from destructive waves and heavy sedimentation in South Florida could accelerate growth to keep pace with sea level rise up to a maximum of 10 mm/year. However, above this rate, reef communities lag sea level and growth is diminished as they become progressively submerged.

In tropical waters within 20° latitude of the equator, atolls form from the remains of reef-forming organisms that accumulate on peaks of submerged mid-ocean volcanoes. Atolls are usually arranged into a rim around a central lagoon. Examination of

Figure 6–14 Stunted mangroves along a tidal creek in the Indus Delta (Pakistan). Decreases in water discharge resulting in loss of sediment will lead to destruction of mangroves from saltwater intrusion and coastal erosion.

reef processes in the Atoll States in the Pacific and the Indian Ocean indicates that in most cases reef growth has lagged behind rising sea level.[92] One of the main controls on equilibrium of atoll islands is storm intensity. Storms generate rubble which, together with coral growth, creates reef flats. A decrease in the frequency of large storms could actually cause an increase in coastal erosion because rare storms of extreme magnitude may be critical in episodic generation of new coral rubble for redistribution. Many reefs, then, will respond to further increases in sea level rise by eroding while simultaneously drowning in place.

The impact of an accelerated sea level rise will not be the same everywhere. One reason is that local erosional processes operate within a broad framework determined by sea level history, which provides an inherited geological character. The amount of erosion will depend on composition and height of a particular island, degree of cementation as determined by rainfall, and on the response to waves and currents. As is the case for other coastal environments discussed in this chapter, high land does not exist. Shoreline and coral rubble elevations are rarely above 3 m on atolls and 1 to 7 m high in large nearshore reef tracts such as the Florida Keys. If sea level rise exceeds the ability of a reef to grow upwards, then storm overwash will increase and inundation will ultimately occur. Time scales for widespread inundation of reefs at present rates of sea level rise (30 to 40 cm/century) is estimated to be on the order of 250 to 500 years.

REFERENCES

1. Davis R. A. and H. E. Clifton. Sea-level change and the preservation potential of wave-dominated and tide-dominated coastal sequences, in *Sea-Level Fluctuation and Coastal Evolution*, D. Nummedal, O. H. Pilkey, and J. D. Howard, Eds. (Tulsa, OK: Society of Economic Paleontologists and Mineralogists, 1987), pp. 166–178.

2. Bird, E. C. F. *Coastline Changes* (New York: John Wiley & Sons, 1985), p. 219.

3. May, S. K., R. Dolan, and B. P. Hayden. Erosion of U.S. shorelines, *EOS* 64:551–552 (1983).

4. Leatherman, S. P. and C. H. Gaunt. National Assessment of beach nourishment requirements associated with accelerated sea level rise, in *Coastal Zone '89*, O. T. Magoon, H. Converse, D. Miner, L. T. Tobin, and D. Clark, Eds. (New York: American Society of Civil Engineers, 1989), pp. 1978–1993.

5. National Research Council. *Responding to Changes in Sea Level* (Washington, D.C.: National Academy Press, 1987), p. 148.

6. Wells, J. T. and J. M. Coleman. Wetland loss and the subdelta life cycle, *Estuarine Coastal Shelf Sci.* 25:111–125 (1987).

7. Gable F. Contemporary climate change and its related effects on global shorelines, in *Coastal Zone '89*, O. T. Magoon, H. Converse, D. Miner, L. T. Tobin, and D. Clark, Eds. (New York: American Society of Civil Engineers, 1989), pp. 1370–1383.

8. Rampino M. and J. E. Sanders. Evolution of the barrier island of southern Long Island, New York, *Sedimentology* 28:37–47 (1981).

9. Penland, S., J. R. Suter, and R. Boyd. Barrier island arcs along abandoned Mississippi River deltas, *Mar. Geol.* 63:197–233 (1985).

10. Pilkey, O. H. and T. W. Davis. An analysis of coastal recession models: North Carolina coast, in *Sea-level Fluctuation and Coastal Evolution*, D. Nummedal, O. H. Pilkey, and J. D. Howard, Eds. (Tulsa, OK: Society of Economic Paleontologists and Mineralogists, 1987), pp. 59–68.

11. Curray, J. R. Transgressions and regressions, in *Papers in Marine Geology, Shepard Commemorative Volume* (New York: Macmillan Press, 1964).

12. Swift, D. J. P. Coastal erosion and transgressive stratigraphy, *J. Geol.* 76:445–456 (1968).

13. Posamentier, H. W., M. T. Jervey, and P. R. Vail. Eustatic controls on clastic deposition I — conceptual framework, in *Sea-Level Changes: An Integrated Approach*, C. K. Wilgus, B. S. Hastings, C. Kendall, H. W. Posamentier, C. A. Ross, and J. C. Van Wagoner, Eds. (Tulsa, OK: Society of Economic Paleontologists and Mineralogists, 1988), pp. 109–124.

14. Posamentier, H. W. and P. R. Vail. Eustatic controls on clastic deposition II — sequence and systems tract models, in *Sea-Level Changes: An Integrated Approach*, C. K. Wilgus, B. S. Hastings, C. Kendall, H. W. Posamentier, C. A. Ross, and J. C. Van Wagoner, Eds. (Tulsa, OK: Society of Economic Paleontologists and Mineralogists, 1988), pp. 125–154.

15. Fairbridge, R. W. Crescendo events in sea-level changes, *J. Coastal Res.* 5:ii-vi (1989).

16. Leatherman, S. P. Coastal Geomorphic responses to sea level rise: Galveston Bay, Texas, in *Greenhouse Effect and Sea Level Rise*, M. C. Barth and J. G. Titus, Eds. (New York: Van Nostrand Reinhold Co., Inc., 1984), pp. 151–178.

17. Byrnes, M. R., R. A. Mc Bride, and M. W. Hiland. Accuracy standards and develop-
    ment of a national shoreline change data base, in *Coastal Sediments '91* (New York:
    American Society of Civil Engineers, 1991), 1027–1042.

18. Morton, R. A. Accurate shoreline mapping: past, present, and future, in *Coastal
    Sediments '91* (New York: American Society of Civil Engineers, 1991), 997–1010.

19. Shore Protection Manual, U.S. Army Corps of Engineers, Coastal Engineering Re-
    search Center (1984).

20. Everts, C. H., J. P. Battley, and P. N. Gibson. Shoreline movements, Cape Henry,
    Virginia, to Cape Hatteras, North Carolina, 1849–1980, USACOE, Technical Report,
    CERC-83-1 (1983), p. 111.

21. Everts, C. H. Sea level rise effects on shoreline position, *J. Port Coastal Waterway
    Eng. ASCE* 111:985–999 (1984).

22. Bruun, P. Sea-level rise as a cause of shore erosion, *J. Waterways Harbors Div., Am.
    Soc. Civ. Eng.* 88:117–130 (1962).

23. Hands, E. B. Observations of barred coastal profiles under the influence of rising water
    levels, eastern Lake Michigan, 1967–71, USACOE, Coastal Eng. Res. Center Tech.
    Rept. No. 76-1 (1976), p. 113.

24. Dean, R. G. and E. M. Maurmeyer. Models for beach profile responses, in *Handbook
    of Coastal Processes and Erosion*, P. D. Komar, Ed. (Boca Raton, FL: CRC Press,
    Inc., 1983), pp. 151–165.

25. Mehta, A. J. and R. M. Cushman, Eds. Workshop on Sea Level Rise and Coastal
    Processes, U.S. Department of Energy, DOE/NBB-0086.

26. Kriebel, D. L. and R. G. Dean. Numerical simulation of time-dependent beach and
    dune erosion, *Coastal Eng.* 9:221–245 (1985).

27. Dolan, R., M. S. Fenster, and S. J. Holme. Temporal analysis of shoreline recession
    and accretion, *J. Coastal Res.* 7:772–744 (1991).

28. Kraft, J. C., M. J. Chrzastowski, D. F. Belnap, M. A. Toscano, and C. H. Fletcher, III.
    The transgressive barrier-lagoon coast of Delaware: morphostratigraphy, sedimentary
    sequences and responses to relative rise in sea level, in *Sea-Level Fluctuation and
    Coastal Evolution,* D. Nummedal, O. H. Pilkey, and J. D. Howard, Eds. (Tulsa, OK:
    Society of Economic Paleontologists and Mineralogists, 1987), pp. 129–143.

29. Leatherman, S. P. Shoreline mapping: a comparison of techniques, *Shore Beach*
    51:28–33 (1983).

30. Dolan, R., B. P. Hayden, and S. May. The reliability of shoreline change measured
    from aerial photographs, *Shore Beach* 48:22–29 (1980).

31. Foster, E. R. and R. J. Savage. Methods of historical shoreline analysis, *Coastal Zone
    '89*, O. T. Magoon, H. Converse, D. Miner, L. T. Tobin, D. Clark, Eds. (New York:
    American Society of Civil Engineers, 1989), pp. 4434–4448.

32. Smith, G. L. and G. A. Zarillo. Calculating long-term shoreline recession rates using
    aerial photographic and beach profiling techniques, *J. Coastal Res.* 6:111–120 (1990).

33. DeLaune, R. D., R. H. Baumann, and J. G. Gosselink. Relationships among vertical
    accretion, coastal submergence, and erosion in a Louisiana Gulf Coast marsh, *J.
    Sediment. Petrol.* 53:147–157 (1983).

34. Stevenson, J. C., L. G. Ward, and M. S. Kearney. Vertical accretion in marshes with
    varying rates of sea level rise, in *Estuarine Variability*, D. A. Wolf, Ed. (Orlando, FL:
    Academic Press, Inc., 1986), pp. 241–259.

35. Turner, R. E. and D. R. Cahoon, Eds. *Causes of Wetland Loss in the Coastal Central Gulf
    of Mexico*, (Minerals Management Service, OCS Study/MMS 87–0119, 1987), p. 536.

36. Mehta, A. J. and R. Philip. Bay superelevation: causes and significance in coastal water level response, University of Florida, Coastal Oceanog. Engr. Dept. Tech. Rept. 061.

37. Fuhrboter, A. and J. Jenson. Long term changes of tidal regime in the German Bight (North Sea), in *Proc. 4th Symp. Coastal Ocean Manage.* (New York: American Society of Civil Engineers, 1985), pp. 1991–2013.

38. Emanuel, K. A. The dependence of hurricane intensity on climate, *Nature* 326:483–485 (1987).

39. Komar, P. D. *Beach Processes and Sedimentation* (Englewood Cliffs, NJ: Prentice-Hall, Inc., 1976), p. 429.

40. Komar, P. D. Beach processes and erosion, in *Handbook of Coastal Processes and Erosion,* P. D. Komar, Ed. (Boca Raton, FL: CRC Press, Inc., 1983), pp. 1–20.

41. Grant, W. D. and O. S. Madsen. Combined wave and current interaction with a rough bottom, *J. Geophys. Res.* 84:1797–1808 (1979).

42. Dyer, K. R. *Coastal and Estuarine Sediment Dyamics* (Chichester, U.K.: John Wiley & Sons, 1986), p. 342.

43. Dean, R. G. Additional sediment input to the nearshore region, *Shore Beach* 55:76–81 (1987).

44. Davis, R. A., W. T. Fox, M. O. Hayes, and J. C. Boothroyd. Comparison of ridge and runnel systems in tidal and nontidal environments, *J. Sediment Petrol.* 42:413–421 (1972).

45. Bruun, P. The Bruun Rule of erosion by sea-level rise: a discussion of large-scale two- and three-dimensional usages, *J. Coastal Res.* 4:627–648 (1988).

46. Fisher, J. J. Shoreline erosion, Rhode Island and North Carolina — tests of Bruun Rule, in *Proc. Per Bruun Symp.* M. L. Schwartz and J. J. Fisher, Eds. (Bellingham, WA: Bureau for Faculty Research, 1980), pp. 32–54.

47. Field, M. F. and D. B. Duane. Post-Pleistocene history of the United States inner continental shelf: significance to origin of barrier islands, *Geol. Soc. Am. Bull.* 87:691–702 (1976).

48. Leatherman, S. P. Migration of Assateague Island, Maryland, by inlet and overwash processes, *Geology* 7:104–107 (1979).

49. Schwartz, M. L. Laboratory study of sea-level rise as a cause of shore erosion, *J. Geol.* 73:528–534 (1965).

50. Schwartz, M. L. The Bruun theory of sea-level rise as a cause of shore erosion, *J. Geol.* 75:76–92 (1967).

51. Hands, E. B. Erosion of the Great Lakes due to changes in water level, in *Handbook of Coastal Processes and Erosion,* P. D. Komar, Ed. (Boca Raton, FL: CRC Press, Inc., 1983), pp. 167–189.

52. Dubois, R. N. Support and refinement of the Bruun Rule on beach erosion, *J. Geol.* 83:651–657 (1975).

53. Dubois, R. N. Nearshore evidence in support of the Bruun Rule on shore erosion, *J. Geol.* 84:485–491 (1976).

54. Dubois, R. N. Predicting beach erosion as a function of rising water level, *J. Geol.* 85:470–476 (1977).

55. Pickrill, R. A. Beach changes on low energy lake shorelines, Lakes Manapouri and Te Anau, New Zealand, *J. Coastal Res.* 1:353–363 (1985).

56. Rosen, P. S. A regional test of the Bruun Rule on shoreline erosion, *Mar. Geol.* 26:M7–M16 (1978).

57. Dubois, R. N. A re-evaluation of Bruun's Rule and supporting evidence, *J. Coastal Res.* 8:618–628 (1992).

58. Bird, E. C. F. World-wide trends in sandy shoreline changes during the past century, *Geog. Phys. Quat.* 35:241–244 (1981).

59. Everts, C. H. Continental shelf evolution in response to a rise in sea level, in *Sea-level Fluctuation and Coastal Evolution*, D. Nummedal, O. H. Pilkey, and J. D. Howard, Eds. (Tulsa, OK: Society of Economic Paleontologists and Mineralogists, 1987), pp. 49–57.

60. Scientific Committee on Ocean Research (SCOR) Working Group. The Response of beaches to sea-level changes: a review of predictive models, *J. Coastal Res.* 7:895–921 (1991).

61. Orford, J. D., R. W. G. Carter, and D. L. Forbes. Gravel barrier migration and sea level rise: some observations from Story Head, Nova Scotia, Canada, *J. Coastal Res.* 7:477–488 (1991).

62. Orford, J. D. and R. W. G. Carter. Crestal overtop and washover sedimentation on a fringing sandy gravel barrier coast, Carnsore Point, Southeast Ireland, *J. Sediment. Petrol.* 52:265–278 (1982).

63. Carter, R. W. G., D. L. Forbes, S. C. Jennings, J. D. Orford, J. Shaw, and R. B. Taylor. Barrier and lagoon coast evolution under differing relative sea-level regimes: examples from Ireland and Nova Scotia, *Mar. Geol.* 88:221–242 (1989).

64. Emery, K. O. and G. G. Kuhn. Erosion of rock shores at La Jolla, California, *Mar. Geol.* 37:197–208 (1980).

65. Sunamura, T. Processes of sea cliff and platform erosion, in *Handbook of Coastal Processes and Erosion*, P. D. Komar, Ed. (Boca Raton, FL: CRC Press, Inc., 1983), pp. 233–265.

66. Doornkamp, J. C. and C. A. M. King. *Numerical Analysis in Geomorphology, An Introduction* (London: Edward Arnold, 1971).

67. King, C. A. M. *Beaches and Coasts* (London: Edward Arnold, 1972).

68. Everts, C. H. Seacliff retreat and coarse sediment yields in southern California, in *Coastal Sediment '91* (New York: American Society of Civil Engineers, 1991), pp. 1586–1598.

69. Wright, L. D. River deltas, in *Coastal Sedimentary Environments*, R. A. Davis, Ed. (New York: Springer-Verlag, 1985), pp. 1–76.

70. Kolb, C. R. and J. R. Van Lopik. Depositional environments of Mississippi River deltaic plain, southeastern Louisiana, in *Deltas in Their Geologic Framework*, M. L. Shirley, Ed. (Houston, TX: Houston Geological Society, 1966), pp. 17–61.

71. Ren, M., R. Zhang, and J. Yang. Sedimentation on the tidal mud flat of China: with special reference to Wanggang Area, Jiangsu Province, in *Proc. Int. Symp. Sediment. Continental Shelf Spec. Reference East China Sea* (Beijing: China Ocean Press, 1983), pp. 1–17.

72. Wells, J. T. and J. M. Coleman. Deltaic morphology and sedimentology, with special reference to the Indus River Delta, in *Marine Geology and Oceanography of Arabian Sea and Coastal Pakistan*, B. U. Haq and J. D. Milliman, Eds. (New York: Van Nostrand Reinhold Co., 1984), pp. 85–100.

73. Academy of Scientific Research and Technology, U.A.R. *Proc. Semin. Nile Delta Sediment.* (Egypt: Alexandria University, 1976), p. 257.

74. Academy of Scientific Research and Technology, U.A.R. *Proc. Semin. Nile Delta Coastal Proc. Spec. Emphasis Hydrodynam. Aspects* (Egypt: Cairo University, 1977), p. 624.

75. Coleman, J. M., H. H. Roberts, S. P. Murray, and M. Salama. Morphology and dynamic sedimentology of the eastern Nile Delta shelf, *Mar. Geol.* 42:301–326 (1981).

76. United Nations Food and Agricultural Organization (UNFAO). Survey of the irrigation potential of the lower Tana River basin, (Canada: Acres International).

77. Klein, G. deV. Intertidal flats and intertidal sand bodies, in *Coastal Sedimentary Environments,* R. A. Davis, Ed. (New York: Springer-Verlag, 1985), pp. 187–224.

78. Wells, J. T. and J. M. Coleman. Periodic mudflat progradation, northeastern coast of South America: a hypothesis, *J. Sediment. Petrol.* 51:1069–1975 (1981).

79. Wells, J. T. and H. H. Roberts. Fluid mud dynamics and shoreline stabilization: Louisiana chenier plain, in *Proc. 17th Int. Coastal Eng. Conf.* (New York: American Society of Civil Engineers, 1981). pp. 1382–1401.

80. Finkelstein, K. and C. S. Hardaway. Late Holocene sedimentation and erosion of estuarine fringing marshes, York River, Virginia, *J. Coastal Res.* 4:447–456 (1988).

81. Kearney, M. S. and J. C. Stevenson. Island land loss and marsh vertical accretion rate evidence for historical sea-level changes in Chesapeake Bay, *J. Coastal Res.* 7:403–415 (1991).

82. Redfield, A. C. Development of a New England salt marsh, *Ecol. Monog.* 42:201–237 (1972).

83. McCaffrey, R. J. and J. Thompson. A record of the accumulation of sediment and t race metals in a Connecticut salt marsh, *Adv. Geophys.* 22:165–236 (1980).

84. Stumpf, R. P. The process of sedimentation on the surface of a salt marsh, *Estuarine Coastal Shelf Sci.* 17:495–508 (1983).

85. Allen, J. R. L. and J. E. Rae. Vertical salt marsh accretion since the Roman Period in the Severn Estuary, southwest Britain, *Mar. Geol.* 83:225–235 (1988).

86. Allen, J. R. L. and K. Pye, Eds. *Saltmarshes.* (Cambridge: Cambridge University Press, 1992), p. 184.

87. Frey, R. W. and P. B. Basan. Coastal salt marshes, in *Coastal Sedimentary Environments,* R. A. Davis, Ed. (New York: Spring-Verlag, 1978).

88. French, J. R. Eustatic and neotectonic controls on salt marsh sedimentation, in *Coastal Sediments '91* (New York: American Society of Civil Engineers, 1991), pp. 1223–1236.

89. Macnae, W. A general account of the fauna and flora of mangrove swamps and forests in the Indo-West-Pacific region, *Adv. Mar. Biol.* 6:73–270 (1968).

90. Woodroffe, C. D. Development of mangrove forests from a geological perspective, in *Tasks for Vegetation Science,* H. J. Teas, Ed. (The Hague: Dr. W. Junk Publishers, 1983), pp. 1–17.

91. Ellison, J. C. and D. R. Stoddard. Mangrove ecosystem collapse during predicted sea-level rise: Holocene analogues and implications, *J. Coastal Res.* 7:151–165 (1991).

92. Roy, P. and J. Connell. Climatic change and the future of Atoll States, *J. Coastal Res.* 7:1057–1075 (1991).

93. Buddemeier, R. W. and S. V. Smith. Coral reef growth in an era of rapidly rising sea level: predictions and suggestions for long-term research, *Coral Reefs* 7:51–56 (1988).

94. Neumann, A. C. and I. Macintyre. Reef response to sea level rise: keep up, catch up, or give up, in *Proc. 5th Int. Coral Reef Symp.* 3:105–118 (1985).

95. Lidz, B. H. and E. A. Shinn. Paleoshorelines, reefs, and a rising sea: South Florida, U.S.A., *J. Coastal Res.* 7:203–229 (1991).

CHAPTER 7

Sea Level Rise: A Worldwide Assessment of Risk and Protection Costs*

Frank M.J. Hoozemans and Cornelis H. Hulsbergen

INTRODUCTION

SOME BACKGROUND

In its first assessment report[1] the Intergovernmental Panel on Climate Change (IPCC) concluded that the global mean temperatures will increase by 0.2 to 0.5°C per decade over the next century if no measures are taken. It is believed that the impact of what is called an "accelerated sea level rise" or ASLR on the coastal zone will be considerable. The IPCC Working Group I (Climatic Change Group) predicts a rise in global sea level ranging from 30 to 110 cm by the year 2100, mainly caused by thermal expansion of the ocean and melting of small mountain glaciers.

Within IPCC Working Group III the Coastal Zone Management Subgroup (CZMS) contributed to the IPCC first assessment report by producing a report on strategies for adaptation to sea level rise,[2] which also included a World Cost Estimate study, WCE.[3] The report defines three ASLR-related response strategies for coastal areas, namely to retreat, to accommodate, or to protect. These strategies were ratified at two successive international workshops held in Miami (1989) and Perth (1990). Each response strategy has different implications for different coastal resources and can be implemented within each national system of integrated coastal zone management planning.

---

* We regret that we could not avoid the use of some abbreviations. We also realize that stumbling upon abbreviations each time is rather boring but we simply had to use them, as many organizations and institutions are involved. By appending a list with the most relevant abbreviations we hope that reading will be facilitated.

0-87371-301-X/95/$0.00+$.50
© 1995 by CRC Press, Inc.

The CZMS developed a common methodology in 1991.[4] This common methodology uses a stepwise approach so that a country is assisted in identifying the actions needed to plan for, and to cope with ASLR impacts. These steps help a country in defining its specific vulnerability; they examine the feasibility of response options at the same time including the country's institutional, economical, technical, and social implications. They also identify needs for assistance, enabling the application of the response options.

The above common methodology primarily aims at application in country case studies, so that each case study produces comparable vulnerability profiles about the identified aspects. Structuring some country case studies in this way facilitated integration of case study results which were made available prior to the UNCED conference in Brazil, in 1992.[5] The evaluation of vulnerability to ASLR is being conducted now in various countries. Studies were either completed, in progress, or in the planning stage in about 30 countries in 1993 by members of the CZMS, an additional 15 cases being sponsored through the UNEP Regional Seas Programme.

To stimulate and coordinate the ongoing work of the IPCC, the World Coast Conference (WCC) was held in The Hague in 1993. The objective of the WCC was primarily to stimulate national and international efforts in support of the development of national integrated coastal zone management plans for low-lying coastal areas. These plans should account for long-term aspects of sea level rise and other effects related to global climatic change. This conference showed that planned development of coastal areas may reduce vulnerability to climatic change, based on results of experiences with the IPCC common methodology. Since the individual country studies could not produce a comprehensive worldwide overview of ASLR-impacts in time for UNCED and WCC, the CZMS initiated a special study to investigate the worldwide impacts of ASLR, adopting the approach laid down in the common methodology: the global vulnerability assessment (GVA).[6]

OBJECTIVES OF THE WORLD COST ESTIMATE

In 1989 the CZMS set high priority to the identification of coastal areas, populations, and resources that would be at risk by ASLR. The Netherlands took the first step in response to this CZMS recommendation, and made the first attempt to cover the issue on an international basis resulting in the WCE report.[4] The objective was to provide a fair indication of the order of magnitude of the coastal defense cost per coastal country and for the whole world. The basic assumption here was a threat of an ASLR of 1 m that had to be dealt with. The study was neither an impact assessment nor did it contain a cost/benefit consideration of impacts and response options. It rather concentrated on costs of relatively simple coastal defense measures that were devised to neutralize the threat of a 1-m ASLR. By doing so the status quo of the safety against flooding should be maintained. As one of the many necessary schematizations, the 1-m ASLR scenario was basically transformed into the general requirement to raise the dikes by 1 m, with some special defense measures for harbor areas and low-lying islands.

OBJECTIVES OF THE GLOBAL VULNERABILITY ASSESSMENT

One of the aims of the CZMS is to provide a worldwide estimate of socio-economic and ecological implications of ASLR, based on the information made available through the case studies, questionnaires, and additional fact finding. However, completion of a sufficient number of IPCC country studies will take several years. For this reason the primary objective of the GVA was set at generating a first overview of the vulnerability of coastal regions on a global scale. Secondary objectives were to provide a reference for IPCC country studies and to investigate the feasibility of some of the steps suggested in the common methodology.

The assessment of risks, losses, and changes for all relevant resources of the world's coastal zones requires detailed worldwide information concentrating on distribution, density, and state of the resources and on the relevant hazardous events and corresponding probability distributions. A problem is that for many resources, such as ecosystems, there are no data available on a global scale. Another constraining factor is that the response of the various systems and their response time scales are not always known at the same level of proven technology, and sometimes not known to a sufficient degree of accuracy at all.

After considering these restrictions it was decided to limit the GVA to impacts on the following three elements in the coastal zone and related areas:

- Population at risk on a worldwide scale
- Coastal wetlands of international importance at loss on a worldwide scale
- Rice production at change in South, Southeast, and East Asia

SCOPE OF THIS STUDY

In this chapter we present the findings of a study of the impacts of an accelerated rise in sea level. However, we have limited ourselves: the outcome focuses only on a selection of quantitative worldwide results of the population at risk in low-lying coastal regions due to ASLR and on an overview of costs of basic protection measures. The study is based on the WCE and the GVA. Impact of climatic change on coastal wetlands and coastal agriculture is discussed in Chapters 7, 10, 11, and 12. In the following the reader may find the backgrounds and the theoretical considerations that formed the basis of the study.

POPULATION AT RISK

CONCEPT OF RISK

According to the definition of *vulnerability* described in the common methodology, vulnerability reflects the degree of capability of a nation to cope with the consequences of ASLR. As such, the concept of vulnerability is involved with a number of different aspects:

- The sensitivity of a coastal area to physical changes imposed by ASLR
- The impacts of physical changes on the socio-economic system and ecological system
- The capability of a region or country to cope with the impacts, including the possibilities to prevent or soothe such impacts by implementation of measures

The present chapter focuses on vulnerability on a regional level, for example, the Asia Indian Ocean Coast, with the aggregated nations' capability to cope with ASLR strictly based on the gross national product.[8]

In order to assess the vulnerability of a coastal zone to ASLR we need uniform procedures to compare and to integrate regional and national studies. For this, the common methodology introduces various procedures. Crucial for these procedures are the concepts of values at risk, values at loss, and values at change. These concepts will help in determining impacts in measurable and objective terms. Of these concepts, we will only consider here the concept of risk for the population. We define the concept of risk as the probability of occurrence of natural hazardous events, multiplied by the consequence of these events. The natural hazardous events may range from worldwide events, such as eustatic sea level rise to regional events like subsidence, changing rainfall, or an increase in storm surges.

The concept of risk, to be elaborated in the section entitled "Definition of Risk", is considered appropriate in the context of assessing the consequences of ASLR for the population in the coastal zone. The short-term consequences of flooding to population vary from minor effects such as flooding of private properties to the possibility of loss of lives. However, future changes in population behavior due to such events are not predicted in terms of the GVA as such a prediction is considered unrealistic. It is, for instance, reasonable in certain instances to assume that frequent flooding may lead to migration, but this is certainly not true for countries like Bangladesh where migration is not a realistic option.

DELINEATION OF THE FLOOD-PRONE AREAS

For the determination of the impacts in terms of risk on the population, we must first define the coastal zone areas in which the risks may occur. In the GVA this zone, which is referred to as *risk zone*, or *flood-prone area*, was defined in terms of areas and accompanying flooding frequencies, including those due to storm surges. The physical consequences of ASLR were derived from consideration of the inundation and flooding process based on the current topography. Corrections for regional and local processes are considered to be less significant insofar as this concerns the population at risk of flooding.

BOUNDARY CONDITIONS AND SCENARIOS

In assessing vulnerability, three types of boundary conditions and scenarios are incorporated in the study, in line with the requirements of the common methodology.

In the first place, the rate of ASLR as the primary external condition imposed by nature should be dealt with. It is true that the common methodology suggests that a rise of, respectively, 0.3 and 1.0 m should be considered by the year 2100 vs. the present situation. However, in this chapter we will only consider the 1.0 m scenario with the remark that the increase in sea level rise will be considered as an instantaneous process. Second, the socio-economic development scenario for the considered coastal regions is being introduced as a main external variable. Since it is impossible to make a reliable prognosis over a period of 100 years, a 30-year population projection is suggested, which was also adopted for this study. Finally, the common methodology introduces the response options for coastal defense as a scenario variable. But here we will adopt only two selected scenarios from the range of response options, namely no measures at all and a standardized protection measure.

The year 1990 was chosen as the present state of the countries to be affected by ASLR. Where recent information about the essential variables was missing, extrapolations were made on the basis of past trends to arrive at 1990 figures. As *future year* for the projected population, 2020 was taken following the guidelines of the common methodology.

DEFINITION OF RISK

ASLR and associated storm surges are natural events harmful to man caused by forces beyond his control. In general, we call these phenomena natural hazards. Natural hazards can best be seen within the framework portrayed in Figure 7–1. In this figure a distinction is made between natural events and their interpretation as natural hazards.

As the earth is a highly dynamic planet, most natural events show a wide variation in intensity. Events occurring at the outer limit of this behavior are called *extremes* and here we use statistical methods to describe them. Extreme natural events are called hazards in case they cause serious damage to life and property. Natural hazards such as ASLR-effects result from the conflict of marine/meteorologic processes with man and the like at the interface of what is commonly called the *natural events system* and the *human user system*. This can be illustrated by means of Figure 7–2. In this figure the shaded area represents an acceptable fluctuation in sea level, bounded by a threshold value representing the socio-economic degree of tolerance. Beyond this limit events are defined as hazards and their magnitude, duration, and probability of occurrence should be considered when assessing the risk associated with these hazards.

As discussed before, risk is defined as the probability of a given event times its effect in terms of affected area, number of people inundated, and economic losses. In general terms risk can be expressed as

$$R = p(I) \times E(I)$$

where: R = risk; I = specific event; p = probability of occurrence of event I; and E = effect of the event I on the socio-economic system.

THE NATURE OF HAZARD

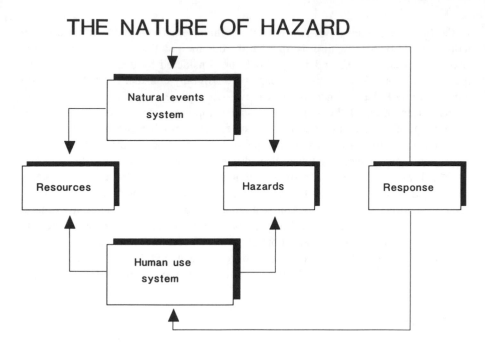

Figure 7–1 Natural hazards occur at the interface between natural events and human use systems. Human response to hazards may modify natural events and human use. (From Smith, K., *Environmental Hazards, Assessing Risk and Reducing Disaster* (London: Routledge, 1992). With permission.)

HAZARD IN THE ENVIRONMENT

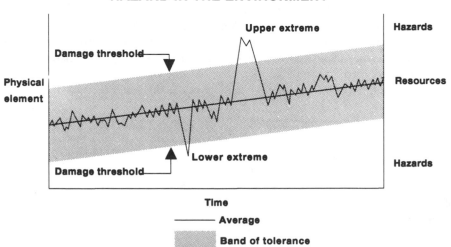

Figure 7–2 Sensitivity to environmental hazard expressed as a function of the variability of geophysical elements and the degree of socio-economic tolerance. Within the band of tolerance, events are perceived as resources; beyond the damage thresholds they are perceived as hazards.

In general E will be a function of space and time. The total risk for a given area can therefore be expressed as

$$R(t) = \sum_{i=1}^{N} p(I_i) \times E(I_i, A_i, t)$$

where: A_i = area where event I_i can happen; N = number of events I_i; and t = time span.

The value of $R(t)$ can be computed for various ASLR scenarios. In this chapter we only regard the present situation and the predicted 1.0-m instantaneous rise in sea level by the year 2100.

EXCLUSION OF RESPONSE FUNCTIONS

When assessing the effect of floods, the response time of the effect under consideration should in general be considered. Phenomena having a large response time, such as ecological responses, act like a low-pass filter, which means that only long periodic fluctuations and trends are of importance. On the other hand, systems for risk assessment with short response times will respond almost instantaneously to changing water levels. This is, for instance, the case when inundation risk of a flood-prone area is considered.

Physical Impact Functions

The physical impact of a flood, defined in terms of inundation depth and inundated area, is determined primarily by the physical characteristics of the flood-prone area. Such an area can be subdivided, for instance, into the inland waters and the river system, estuaries and lagoons, and the coastal plain and the coastal strip. The intensity and duration of the inundation in a particular area can be computed by means of hydrodynamic models for surface and river flow.

Socio-Economic Impact Function

The effect of flooding can be assessed in different ways, using increasing levels of sophistication in the description of the impact function $E(I)$. At the basic level $E(I,A)$ is merely defined as the average population at risk. This estimate can further be fine-tuned by introducing the population density as a function of I and A. A more realistic estimate is obtained when the area at risk is defined as the area with inundation depths of more than 1 m as threshold value for the risk of drowning. In this chapter only the first method has been used to estimate the population at risk and no response function has been taken into account.

METHODOLOGY

Approach and Assumptions

As a general approach, the following steps were undertaken in the GVA to determine the population at risk for the various scenarios:

- Assessment of the height of the maximum flood level theoretically threatening the low-lying coastal zone, taking into account present and possible future extreme hydraulic and geophysical conditions
- Determination of the flood-prone area and calculation of the area captured between the coastline and the maximum flood level
- Assessment of the present state of protection against flooding in the low-lying countries
- Determination of the population densities for the present and future state on a subnational basis
- Determination of the population at risk with and without measures, with and without ASLR, for present and for future conditions

Each of these steps is addressed in the following sections, but first we will discuss some of the limitations, assumptions, and associated effects.

Limitations of Data Sources

The worldwide data sources that were used allowed only for a limited spatial resolution of a number of variables like population distribution, design water levels, and surface areas between design water levels. This introduces inaccuracies and requires assumptions, all leading to inaccurate results on the scale of countries themselves. Verification has shown that averaged results on regional scales are rather accurate.

No Physical System Response

In the assessment of population at risk any physical changes in coastal environment in the course of time are ignored. Although in many cases this seems to be a realistic assumption because of the artificial restrictions of sediment availability by river regulation, river damming, and coastal protection, this assumption is not generally true. The effects of these changes on the population at risk are, however, estimated to be less significant.

Hydraulic Conditions and Regional Effects

It was necessary to make assumptions about effects other than increasing storm surge levels such as subsidence, tides, and barometric pressure. These estimates again limit the accuracy of results on a country scale but less so on a regional scale.

Estimate of Present Protection Status

Since there is no information available on a global scale about the present protection status of the world's coastlines, it is assumed that the estimate could be based on the per capita GNP; this is discussed in the section entitled "Assessment of the Present and Future Protection Status".

Assessment of Hydraulic and Geophysical Conditions

Essential in each assessment of impacts caused by ASLR is the determination of risk zones related to hydraulic and geophysical conditions. Low-lying coastal zones are prone to flooding when they are lying below a specified flood level. This flood level is determined by a number of regional factors:

- *Global hydraulic factors* — The global hydraulic factor accounted for in the GVA study is the ASLR of 1 m/100 years.
- *Regional geophysical factors* — Subsidence and tectonic uplift are accounted for, but based on crude information.
- *Regional hydraulic factors*
  - Tides (mean high water spring)
  - Storm surges (up to an exceedance frequency of once in 1000 years)
  - Barometric pressure

In the following we shall deal with each condition in more detail.

Subsidence

In the calculation of the risk zones, attention has been paid to the phenomenon of subsidence until the year 2020.[9,10] In general, one single figure for each country cannot be given for such a strictly local phenomenon. However, it was found that most of the low-lying coastal areas are situated in deltaic areas, each with more or less similar tectonics. For these areas more information is available although still rather superficial. The local information is being used to represent the geotectonics for the whole coast. The overestimate with regard to the population at risk finally made is rather small because: (1) the subsidence and uplift estimates are very conservative, and (2) the number of people living in low-lying areas outside subsiding deltas is relatively small.

Tides

As to the global tidal data, mean high-water levels derived from the Admiralty Tide Tables[14] were used. For each country one tidal level (mean high-water level) was calculated based on information of measurements in standard and secondary ports. For some countries the coastal zone had to be split up into various regions.

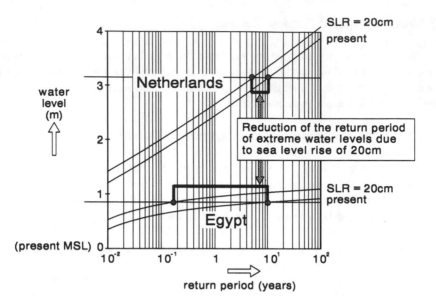

Figure 7–3 The effect of a 0.20-m SLR on the return period of storm surge levels.

Storm Surges

For the estimation of the storm surges at the coasts of all the countries of the world, the following steps can be recognized:

- Estimation of the wave climate in the sea area off the coast
- Extrapolation to long return periods
- Computation of decisive wind speeds
- Determination of the characteristic depth and bottom slope of the coast
- Computation of storm surge

It should be noted that characterizing the hydraulic conditions of an entire country results in a very rough spatial averaging of storm surges.[11-13] High extreme conditions are usually encountered in rather small parts of the coast like estuaries, but the error with respect to the total area at risk that is made by averaging is expected to be small. Because of worldwide differences in the regime of tides and storm surge events, the increase in flood risk due to ASLR will be larger than average for some coastal regions and lower for others. An illustration of differences in storm surge regime and of the impact of ASLR is given in Figure 7–3. The curves are smoothed and extrapolated in the range above the 100-year return period, but otherwise based on observed storm surge levels. As indicated in this figure, a sea level rise of 0.2 m, for example, will decrease the return period of a fixed design water level. Note, however, that in coastal regions characterized by a mild curve slope, e.g., occurring in Egypt, the return period of extreme water levels will decrease more rapidly for a given increase in sea level than in coastal regions with a steep curve like the Netherlands. The differences in steepness are related to the nearby shelf geometry and to the regime of extreme meteorological conditions.

Barometric Pressure

With respect to the barometric pressure, no information on a regional scale is available and for this reason this factor is dealt with as a global hydraulic factor. Based upon the assumption that the passage of a storm will result in a local, temporary heightening of the water level, several storm surge scenarios are adopted for changes in the atmospheric pressure.

Determination of the Flood-Prone Area

By making an overlay of the maximum flood level based on storm surges with an exceedance frequency of once in a thousand years and data about the geographical features (elevation) of the low-lying coastal area, the surface area of the potential risk zone or flood-prone area has been calculated.[15] For this risk zone of the coastal strip it may be expected that the number of inhabitants living in this area will be affected in some way or another by an increase in sea level. As a basis for the determination of the flood-prone area we used two global databases:

1. *The ETOPO-5 Elevation database* — acquired from UNEP/GRID-PAC, Geneva.[16] This database comes originally from the U.S. National Geophysical Data Center in Boulder, Colorado. It has elevation readings sampled from every 5-min latitude/ longitude crossing on the global grid (approximately 80 km² spatial resolution, or 12 × 12 pixels/degree), and a 1-m contour interval. Some original data sources include the U.S. Defense Mapping Agency for the U.S., Japan, and Western Europe, the Australian Bureau of Mineral Resources, and the New Zealand Department of Scientific and Industrial Research.
2. *The World Boundary Database-II Natural and Political Boundaries data sets* — acquired from UNEP/GRID-PAC in combination with boundary data recently prepared by GRID-Geneva.

Assessment of the Present and Future Protection Status

When dealing with the notion of people at risk, the probability of flooding events is a notion of crucial importance. The probability (frequency) of flooding events may typically be reduced, for a given area susceptible to flooding, by improving the protection status for that particular area.

The basic idea is that raising the protection status by an investment in protection improvement (e.g., by constructing a higher artificial barrier between the sea and the occasionally flooded land) reduces the probability of flooding events. In historic empirical practice dikes were built and subsequently raised, the height of the calamitous flood level being used as the new dike crest elevation target (design height), to the extent that communities could afford. Only recently has the local crest design height been scientifically based both upon water level probability and cost/benefit consider-ations.

Because of this highly empirical background, the present protection status of the world's coastlines is in most cases not well known. Therefore, we had to make an assumption to estimate the protection status of all coastal nations of the world. We assumed a general congruence between the country's economic development level and its protection status. We expressed the protection status not in terms of a crest elevation in meters above some sea level, but in terms of the design exceedance frequency of the water levels, and the development level in terms of its gross national product (GNP per capita) (see the table below).

The design exceedance frequency with respect to a protective structure (a dike) is that water level exceedance frequency for which the dike under design will just withstand the attack; for higher water levels (with a lower frequency than that used in the design) flooding will occur. It may be clear that for every single location and situation a particular, highly site-specific curve exists (although unknown to us until appropriate measurements and analysis are carried out), describing the relation between each water level and its exceedance frequency. For example, a natural, low-lying coastal area without artificial protection may be flooded during every major storm surge, say ten times per year on average. The protection status of this area may be defined with a design exceedance frequency of ten times per year, or once in 0.1 year. Protecting this area with a low dike will result in a decrease in the average flooding frequency. The new protection status will be expressed as, e.g., once per year. Further development of that low-lying area will undoubtedly result in higher dikes, causing further reduction in the flooding. In this chapter the future coastal protection is assumed to be as follows:

- The protection measures will be restricted to the implementation of a coastal defense system of a higher protection class by one step: PC-1 is being increased to PC-2, and PC-2 to PC-3. A protection measure where the lowest protection class (PC-1) will be heightened to the safest protection class (PC-3) is not expected to be realistic.
- Protection class PC-3 is considered the maximum protection level and will not be improved. Some countries such as the Netherlands may even adopt higher protec-tion levels (design frequency up to 1/10,000 years), but such cases are rather exceptional and can be neglected within the global scale of the GVA.

The following table summarizes the above-described methodology.

GNP in U.S. $ per capita (1990)	Present situation		Future situation (with measures)	
	Protection class (PC)	Design frequency	Protection class	Design frequency
< 600	PC-1	1/10 years	PC-2	1/100 years
600–2400	PC-2	1/100 years	PC-3	1/1000 years
>2400	PC-3	1/1000 years	PC-3	1/1000 years

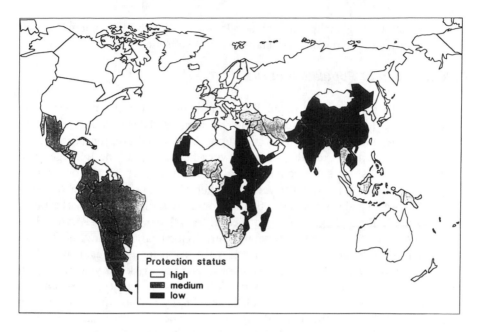

Figure 7–4 Assumed protection status of the world's coasts.

Figure 7–4 presents a worldwide overview of the protection status in 1990 derived from the above-mentioned analysis.

Determination of Population

The common methodology requires two population density scenarios for low-lying coastal areas: the present (1990) one and a future population density (2020). As no worldwide information is available on population living in the low-lying coastal areas, it is assumed to be equally distributed over the overall low-lying coastal province or district. In order to account for regional differences in population density, the population density per province or district of each country was estimated based on country statistics.

The regional population densities were collected mainly from general sources as the *Encyclopedia Britannica* or from selected country data:[19-23]

- For all the countries of the world there are population estimates of the year 1990 on a national scale but estimates are not always available on a regional level. In order to obtain the 1990 regional population data, historical regional population data has been updated to the 1990 level based upon the trend of the regional population development and the 1990 national total.
- If no data about the trend of the regional population development is available, the national population growth figure was applied to the most recent regional population data available. The same procedure has been followed for the calculation of the regional population in 2020. Both the World Bank and UN provide frequent updates of national population projections.

Using both the regional population development trend and the national growth figure, estimates of the regional populations could be made.

Determination of Population at Risk

In order to calculate the population at risk in the various risk zones an assumption should be made of how people and compound design water levels are distributed over the elevation levels.[24-26] Suppose that near a region of interest a water level of +4 m is exceeded once in every hundred years on average ($f_3 = 0.01$/year). The exceedance frequency of a water level of +6 m will be once in every thousand years ($f_4 = 0.001$/year). Furthermore, suppose that in the area between the +4-m and +6-m lines 100,000 people are living ($POP_{3,4} = 100,000$). In this example the average number of people at risk per year (PAR) would be equal to $PAR_{3,4} = 1000$ provided all people were living near the +4-m line. On the other hand, if all the people were living near the +6 m-line, the result would be $PAR_{3,4} = 100$. In reality, the average number of people at risk per year will lie in between these two limits. In the GVA it has been assumed that the exceedance frequency for level h_i is given by f_i and for level h_j by f_j. The number of people living between these heights is POP_{ij}. Here $POP(f)$ denotes the total population living below the level with exceedance frequency f. As the exceedance frequency is inversely proportional to an exponent of the height and people are assumed to live evenly dispersed over all heights, the population volume $POP(f)$ is linearly related to $\log(f)$. Since in the GVA the frequencies of exceedance are expressed in decades, that is $f_i = 10 \times f_j$, the outcome of the above schematization results in the general equation:

$$PAR_{i,j} = \frac{9 f_j}{\log \left(f_f / f_i \right)} POP_{i,j} = 3.9 f_j POP_{i,j}$$

For the above example, the best estimate for the average population at risk per year between the +4 and +6-m lines would be 390. By making use of this equation, the total population at risk per year per country can be determined for each protection situation.

COASTAL PROTECTION AND COSTS

GENERAL APPROACH

To cope with the extremely wide range of existing sorts of coastal defense measures and their related costs within the constraints of the present chapter a modular approach was adopted. The following modules were taken into account:

- Lengths of the coastline (in some cases, coastal areas), to be protected by various standardized coastal defense measures
- A limited set of six standardized coastal defense measures
- Reference unit costs for each type of standardized coastal defense measure
- Individual country cost factors

For each individual country the national costs were found by applying the following formula:

Sum of [coastline length × defense measure cost per unit of length × country cost factor] for all partial stretches of coastline in that country or coastal area which need protection

Regionally and worldwide, the aggregated costs are found by summing over the respective national costs. However, this schematic setup should not be misinterpreted as it does not produce a basis for national coastal defense planning. But the modular setup as such forms a realistic and most practical framework for subsequent, more detailed analyses to improve the accuracy and local relevance of the individual modules.

LENGTHS OF COASTLINE TO BE PROTECTED BY STANDARDIZED COASTAL DEFENSE MEASURES

For all countries an evaluation was made of the present type of sea defense works. Assuming that new structures have to be made the preferred type of defense can be selected based on this information. This selection should also account for matters like soil conditions, height and sea-exposure of the shore, the availability of construction materials, and the value of the direct hinterland.

As the GVA does not assess the exact length of the coastline to be protected, the coastline length of the WCE study has been used, adding the length of the low coast, the length of the city waterfronts, and the length of the low coast of islands (Annex D2 and D3 of the WCE study).[3] Because the composition of the coastline may differ considerably within a country (dunes, dikes, marshes) and because no global database about the composition of the coast is available, a choice was made for a typical coastline per country. This selection was based on the most dominant coastline in the particular country. This procedure is similar to the one adopted in the WCE study. Since also in the WCE study information about the composition of the coastline was lacking, a typical "standard" defense measure was proposed for each country.

DEFINITION OF STANDARD PROTECTION MEASURES

For the protection of people living in low-lying coastal areas the standard protection measures 1, 2, and 3 apply in most cases.

1. *Stone-protected sea dike*
   Dimensions   height — variable
                            crest width — 5 m
                            slopes — 1:2.5
                            bottom level — at HHWS
                            toe of slope — below LLWS
   Construction   core of locally available material
                            filter layer of 0.5 m of gravel/stone
                            stone layer 1.5 m of medium-sized stone
                            armour layer 2.5 m of large-sized stone/blocks
   To be applied   on lower erodible shores
                              where severe wave attack is prevailing

2. *Clay-covered sea dike*
   Dimensions   height — variable
                            crest width — 5 m
                            slopes — 1:5 outer
                                   — 1:3 inner
                            bottom level — at HHWS
   Construction   core of local available material
                            cover layer 1 m of clay with grass vegetation
   To be applied   on lower nonerodible shores where wave attack is
                                reduced
                              where land use and coastal formation allows for
                                this type of defense

3. *Sand dune*
   Dimensions   height — variable
                            crest width — 150 m
                            slopes — 1:4 outer
                                   — 1:3 inner
                            bottom level — at HHWS
   Construction   body of reclaimed sand, mined from offshore sources
                            protected by natural vegetation
   To be applied   on lower sandy beaches where sand is available for
                                construction
                              where sufficient land is available

   In first instance type 2 (clay-bund) can be selected. Only where the greater part of the coastline is formed by dunes, type 3 (dune) should be applied. Type 1 (dike) is only selected in situations where the shoreline is presently also firmly protected by stone dikes, mainly in densely populated areas. Where no clear selection could be made, the cheapest solution (Type 2) has been applied.

4. *Tourist beach maintenance* On those coastlines where important tourist beaches must be maintained, beach nourishment schemes will be necessary. This measure does not follow from safety considerations: on low sandy coasts where safety is at stake, standard protection measure No. 3 applies. The following standard measure holds:

Dimensions: 1 m thick sand layer across morphologically active zone, up to MSL 8-m depth line, approximately a 1000 m wide

Construction: nourishment by trailing suction hopper dredge from offshore sources, or by cutter suction dredge from near-shore sources

To be applied: to maintain important tourist resort beaches

5. *Harbors and industrial areas* For low-lying harbors and associated low-lying industrial areas the following standard protection measure applies:

Construction: low-lying outer dike areas will be raised by 1 m and quay walls will be strengthened

To be applied: low-lying harbors and associated low-lying industrial areas

6. *Elevation of low-lying small islands* Small, very-low-lying (coral) islands require special measures since water can easily penetrate into the subsoil. This, in turn, excludes the application of dikes with additional pump drainage. These very-low-lying islands lie only 0.5 to 1.5 m above mean sea level.

Construction: existing coral islands will be raised by 1 m by placing material from offshore sources or from other islands that must be abandoned for that purpose.

To be applied: on small, very-low-lying (coral) islands.

The above standardized measures are believed to be realistic approaches for most countries of the world. However, execution of such measures requires a well-functioning technical and organizational infrastructure. More demanding coastal defense works like those used to close off large estuaries were deliberately omitted from the standard list, although their application may be the most economical solution in special cases.

UNIT COSTS

Cost estimates for the standard protection measures are established based on the following assumptions and conditions:

- Standard defense constructions are defined for various distinct situations. The definitions include dimensions, construction material, and construction methods.
- The schematized designs are based on Dutch procedures as developed during centuries of ongoing combat against flooding, in combination with experiences obtained during numerous construction works abroad.
- Construction method and cost estimates are based on the assumption of construction in one continuous operation per project. Cost effects (positive or negative) of a gradual implementation of the works over a longer period are ignored. Also, quantum effects by a combined execution of various smaller projects are not accounted for.
- Cost estimates are based on the price level of the end of 1990.

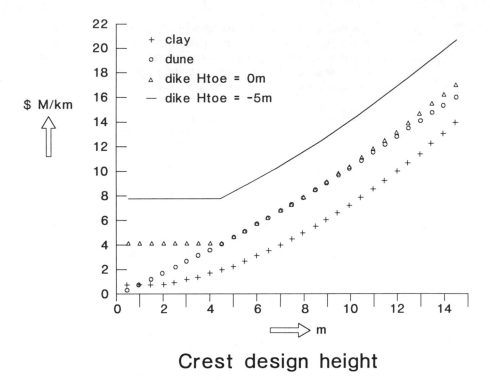

Crest design height

Figure 7–5 Construction costs for different protection measures.

The following unit costs apply, based on the Dutch situation:

No.	Standard measure Type	Unit costs	Remarks
1	Stone dike	8.5 M$/km	If toe is 5 m below HHWS
	Stone dike	4.5 M$/km	If toe is 0 m below HHWS
2	Clay dike	2.5 M$/km	
3	Sand dune	4.5 M$/km	
4	Tourist beach	4.5 M$/km	
5	Harbors	15 M$/km²	
6	Low island elevation	12.5 M$/km²	

Figure 7–5 presents a diagram referring to standard measures 1, 2, and 3. For crest design heights that differ from the standard dimensions, one can roughly assess the relevant unit costs by using this diagram. Needless to say, the diagram should not be used for purposes other than those this chapter is dealing with, which also goes for all other data in this chapter.

COUNTRY COST FACTOR

Assessment of worldwide construction costs is obtained by multiplying the Dutch unit costs with a country cost factor. This factor accounts for local circumstances, such as:

Table 7–1 Impact of ASLR on Global Population

| | | 1 Meter ASLR | | |
| | | No defense measures | | With defense measures | |
		1990	2020	1990	2020
World population [million]	5100	5100	7600	5100	7600
Population in risk zone [million]	210	260	400	260	400
People at risk [million/year]	47	61	100	7	12

- The presence of a "wet" civil construction industry
- The availability and costs of human resources
- The availability and quality of construction material
- The possibility for employment of (foreign) equipment
- Possible side effects
- (Land-) acquisition cost
- The local market situation

After determining the population at risk the effects of certain protection measures on the population at risk have been analyzed. For this analysis an identical procedure was used as described above except for an improved protection status: if a certain country had a protection status PC-1 on the basis of its GNP per capita the protection status would be increased to PC-2.

PRESENTATION OF RESULTS ON A GLOBAL AND REGIONAL LEVEL

POPULATION AT RISK

First, worldwide results of the analysis of people at risk are presented in Table 7–1. It shows that the population living in flood-prone coastal areas will increase from 210 million to 260 million due to a sudden ASLR, and to 400 million by the year 2020 due to population growth. Without taking any measures, the population at risk per year will increase from 47 million to 61 million due to a sudden ASLR, increasing to a 100 million by the year 2020. If the defense measures discussed earlier are applied, the population at risk per year would reduce to a level of 7 million per year for the reference situation (1990), increasing to 12 million per year for 2020.

Since one of the goals of the GVA was to identify the relative vulnerability of the various regions in the world to ASLR, we now analyze the regional vulnerability based on a subdivision of the world coasts into 20 coastal regions. These 20 regions are presented in Table 7–2, which relates to the situation of 1990, with ASLR, without defense measures. This table shows the absolute numbers of people in the risk zone and people at risk per year, and expressed as a percentage of the total population in the coastal countries. The coastal regions with the highest relative number of people living in the risk zone are the Southeast Asia region, North and West Europe, the small island states in the Indian and Pacific Oceans, and the Caribbean Islands. Next, looking to the

Table 7–2 Risk Zones and People at Risk per Year by Coastal Region, in 1990, Without Defence Measures

Coastal regions	1990 Population (× 1000)	Population in risk zone (with ASLR, no defence measures) (× 1000)	% of regional popul.	vs. world average	People at risk per year (with ASLR, without defence measures) (× 1000/ year)	%of regional popul.	vs. world average
North America	360,000	16,900	4.7	0	170	0.0	—
Central America	28,500	1,160	4.1	–	56	0.2	—
Caribbean Islands	32,900	2,600	7.9	++	110	0.3	—
S. American Atl. Coast	238,000	6,000	2.5	—	410	0.2	—
S. American Pac. Coast	46,300	2,000	4.3	–	100	0.2	—
Atl. Ocean small isl.	580	20	3.4	–	0	0.0	—
North and West Europe	281,000	24,600	8.8	++	130	0.0	—
Baltic States	51,700	660	1.3	—	3	0.0	—
N. Mediterranean	127,000	5,900	4.6	0	37	0.0	—
S. Mediterranean	171,000	9,400	5.5	0	2,100	1.2	0
African Atl. Coast	267,000	9,900	3.7	–	2,000	0.7	–
African Indian Oc. Coast	193,00	8,500	4.4	–	3,600	1.9	++
Gulf States	95,200	540	0.6	—	14	0.0	—
Asian Indian Ocean Coast	1,159,000	64,100	5.5	0	27,360	2.4	++
Indian Ocean small isl.	2,500	290	11.6	+++	100	4.0	+++
Southeast Asia	394,000	35,500	9.0	++	7,800	2.0	++
East Asia	1,349,000	61,700	4.6	–	17,100	1.3	0
Pacific large islands	24,100	790	3.3	–	17	0.1	—
Pacific small islands	2,300	180	7.8	++	34	1.5	+
C.I.S. (former USSR)	290,000	10,210	3.5	–	52	0.0	—
World	5,100,000	260,900	5.1		61,300	1.2	

Legend regarding relative regional risk level: —, less than 50% of world average; –, 50 to 90% world average; 0, 90 to 110% world average; +, 110 to 150% world average; ++, 150 to 225% world average; +++, more than 225% of average.

relative size of people at risk per year we find that North and West Europe and the Caribbean are less vulnerable, despite the high number of people living in the impact zone compared to the world average. Also vulnerable, based on the relative population at risk per year, are the Indian Ocean coasts of Africa and Asia.

An indication of regional vulnerability can be obtained by dividing the population at risk per year by the total regional population. In this respect, small island states in the Indian Ocean are the most vulnerable, followed by the Asian countries bordering the Indian Ocean, the African Indian Ocean states, and the small island states in the Pacific Ocean. The countries on the southern border of the Mediterranean show an average vulnerability to ASLR.

The same results may also be considered from a different view. By dividing the number of people in the risk zone by the population at risk per year a safety index can be derived for the people that are living in the flood-prone area, as shown in Table 7–3. The higher the safety index, the longer the average return period of a flooding event affecting the population. The safety index per country depends on the assumed protection level. By combining countries with different protection levels, the average safety index of each coastal region is shown. Table 7–3 shows that the countries in the developing

Table 7–3 Saftey Index of Population in Risk Zone

	No measures		With measures	
Coastal regions	1990	1990	1990	1990
North America	99	+	194	+++
Central America	21	0	191	+++
Caribbean Islands	24	0	155	+++
S. American Atl. Coast	15	–	126	++
S. American Pac. Coast	20	0	188	+++
Atl. Ocean small isl.	100	+	82	+
North and West Europe	189	+++	193	+++
Baltic States	194	+++	192	+++
N. Mediterranean	162	+++	188	+++
S. Mediterranean	4	—	37	+
African Atl. Coast	5	—	45	+
African Indian Oc. Coast	2	—	22	0
Gulf States	39	+	194	+++
Asian Indian Ocean Coast	2	—	21	0
Indian Ocean small isl.	3	—	25	+
Southeast Asia	5	—	40	+
East Asia	4	—	28	+
Pacific large islands	45	+	191	+++
Pacific small islands	5	–	51	+
C.I.S. (former USSR)	195	+++	195	+++
World	4.3		35	

Legend: —, index less than 5; –, index between 5 and 20; 0, index around 20; +, index between 20 and 100; ++, index between 100 and 150; +++, index more than 150.

world presently show a very low safety index in line with their GNP per capita. With measures the safety level increases all over the world to more acceptable levels.

In Table 7–4 we show the increase in world population and people at risk per year in 2020. Without measures, the global increase in population at risk per year since 1990 is 63%, which is higher than the increase in world population of 49%.

The highest growth rates of people at risk are found along the coasts of the African Atlantic and Indian Oceans and in the Gulf States. The growth in the latter region is not very essential in terms of absolute numbers. The situation in Africa, however, is more serious. Here, the growth rate of the population at risk in the coastal zone is clearly the highest in the world.

In absolute figures, the Asian Indian Ocean Coast region is the most vulnerable with almost half of the entire world growth of population at risk.

COASTAL PROTECTION EFFECTS AND COSTS

In Table 7–5 the effects of coastal defense measures are presented in terms of a decreasing number of people at risk and the protection cost involved. In the WCE study it was found that the total cost would be around U.S. $500 billion, but according to the GVA calculations costs would amount to about U.S. $1000 billion if hydraulic conditions are accounted for. Clearly, the protection costs as assessed in

Table 7–4 Increase in World Population and People at Risk per Year in Coastal Regions by 2020, With ASLR, Without Measures

Coastal regions	World population in 2020 (× 1000)	Increase vs. 1990 (%)	Regional population at risk/year in 2020 (× 1000/year)	Increase vs. 1990 (%)	Increase vs. world average
North America	462,000	28	320	85	+
Central America	54,500	91	110	94	+
Caribbean Islands	45,900	40	160	42	–
S. American Atl. Coast	347,000	46	560	38	–
S. American Pac. Coast	70,000	51	160	58	–
Atl. Ocean small isl.	1,100	90	0	106	++
North and West Europe	280,000	0	130	1	—
Baltic States	56,900	10	4	3	—
N. Mediterranean	131,000	3	40	5	—
S. Mediterranean	306,000	79	3,500	64	0
African Atl. Coast	549,000	106	5,050	156	+++
African Indian Oc. Coast	436,000	126	8,100	125	++
Gulf States	208,000	118	40	185	+++
Asian Indian Ocean Coast	1,938,000	67	46,500	70	+
Indian Ocean small isl.	4,200	68	240	131	++
Southeast Asia	607,000	54	13,500	73	+
East Asia	1,736,000	29	21,400	25	—
Pacific large islands	33,700	40	36	104	++
Pacific small islands	3,900	70	60	74	+
C.I.S. (former USSR)	334,000	15	60	10	—
World	7,610,000	49	100,100	63	

Legend for percentage increase in population at risk per year: —, less than 30%; –, between 30 and 60%; 0, between 60 and 70%; +, between 70 and 100%; ++, between 100 and 150%; +++, more than 150%.

this study underestimate the total costs of ASLR since they only focus on the costs of protection against the inundation effects of ASLR. No costs have been taken into account for:

- Retreat measures
- Operational water resources management
- Adaptations of the water resources management infrastructure
- Remaining losses

Various case studies carried out so far indicate that costs in the water resources management sector may be of the same order of magnitude as those for the coastal protection. Costs that are very hard to avoid are called "remaining losses". These costs may, e.g., occur as a result of changes in the ecosystem in the coastal zone.

Table 7–5 shows that based on the reduction of the population at risk per year by improvement of coastal defense systems and the annual protection costs in percentage of the GNP, regions like the Asian Indian Ocean Coast, East Asia, the South American Pacific Coast, and the South Mediterranean are considerably effective in terms of risk reduction.

Table 7–5 Costs of Measures and Reduction of People at Risk in Coastal Regions

Coastal regions	Decrease people/ GNP	Low coasts	Harbor elev.	Isl. elev.	Beach nour.	Total costs	Costs per	Annual cost (× 1000)
North America	90	101,900	5,700	0	29,700	137,300	380	0.02%
Central America	49	6,100	110	0	530	6,740	240	0.23%
Caribbean Islands	100	5,600	2,300	110	4,600	12,610	380	0.21%
S. American Atl. Coast	340	119,000	2,700	0	1,600	123,300	520	0.25%
S. American Pac. Coast	108	270	210	0	110	590	10	0.01%
Atl. Ocean small isl.	0	40	0	0	110	150	260	0.07%
North and West Europe	0	70,000	5,800	0	15,400	91,200	320	0.02%
Baltic States	0	36,000	1,200	0	3,000	40,200	780	0.08%
N. Mediterranean	5	8,900	2,300	0	14,300	25,500	200	0.02%
S. Mediterranean	1,890	12,400	3,300	0	2,400	18,100	110	0.07%
African Atl. Coast	1,760	39,000	1,300	0	1,300	41,600	160	0.25%
African Ind. Oc. Coast	3,220	39,000	950	0	1,500	41,500	210	0.38%
Gulf States	11	9,900	3,700	0	210	13,810	150	0.05%
Asian Indian Oc. Coast	24,320	172,000	740	0	2,000	174,740	150	0.52%
Indian Oc. small isl.	90	930	0	1,680	630	3,240	1,300	0.72%
Southeast Asia	6,940	62,000	1,600	0	2,800	66,400	170	0.20%
East Asia	14,950	133,000	8,300	0	3,300	144,600	110	0.06%
Pacific large islands	13	35,000	1,900	0	4,300	41,200	1,710	0.17%
Pacific small islands	30	1,440	110	2,210	530	4,290	1,870	0.77%
C.I.S. (former USSR)	0	58,000	1,100	0	4,000	63,100	220	0.02%
World	53,890	910,000	43,000	4,000	92,000	1,050,000	210	0.056%

THE ASSESSMENT RECAPITULATED

In this chapter we have dealt with one major impact of the acceleration in the rate of sea level rise that is expected due to climate change: the increased flooding risk for the growing population living in the low-lying areas of the world, and the costs necessary to mitigate this risk. The distribution over the world of the threats and the costs has been considered in terms of some 20 regions. At present, without acceleration of the sea level rise, some 200 million people live in coastal areas that are lower than the once per thousand years storm surge level. This number will be more than doubled in 2020 due to ASLR and population dynamics. From the regions considered, especially the coastal populations of the Asian Indian Ocean Coasts, the South Mediterranean Coasts, the African Atlantic as well as Indian Ocean Coasts, and the coasts of many small island states are threatened.

The worldwide costs for just the basically essential protection works are estimated at U.S. $1000 billion. In some regions cost forms a very substantial part of the GNP, especially in the Indian Ocean and Pacific Ocean small island states. When the "effectivity and affordability" of the protection is considered in terms of the number of people that are protected from flooding in relation to the protection cost as a fraction of the GNP, it appears that South and Southeast Asia, the South American Pacific Coast, and the South Mediterranean are much easier to protect than many other regions.

In this global assessment, remedial actions to reduce impacts of ASLR focus on relatively low-technology measures, the emphasis being on dike raising. Closure of river mouths and estuaries by (storm-surge) barriers were not accounted for although the outcome of a cost-benefit analysis for a particular area may indicate such a construction. Based on the wide experience in the Netherlands with the flexible closure of former estuaries by means of storm-surge barriers, dams, and sluices we conclude that the implementation of high-technology hydraulic infrastructure is extremely expensive. Also high-technology solutions invoke high maintenance costs. These dedicated measures, therefore, are beyond GVA scope.

Finally, it is emphasized that the present analysis, based on a large number of assumptions and estimates, must be seen only as a first, global indication of how the pain of accelerated sea level rise will affect the various coastal regions of the world. Many important aspects will only become clearer if site-specific studies are undertaken. As a trigger for such studies, the rapid autonomous development and population increase of the coastal zones will be just as important as the expected accelerated rate of sea level rise.

ABBREVIATIONS

ASLR   — Accelerated Sea Level Rise
CZMS   — Coastal Zone Management Subgroup of IPCC Working Group III
GNP   — Gross National Product
GVA   — Global Vulnerability Assessment
HHWS   — Higher High Water Spring
IPCC   — Intergovernmental Panel on Climate Change
LLWS   — Lower Low Water Spring
PAR   — Population At Risk
PC   — Protection Class
POP   — Population
SLR   — Sea Level Rise
UNCED — United Nations Conference on Environment and Development
UNEP   — United Nations Environment Programme
WCC   — World Coast Conference
WCE   — World Cost Estimate

REFERENCES

1. IPCC. *First Assessment Report* (Geneva, 1990).
2. IPCC. *Strategies for Adaption to Sea Level Rise,* Report of the IPCC Coastal Zone Management Subgroup, (The Hague: Rijkswaterstaat, Ministry of Transport and Public Works, 1990).

3. DELFT HYDRAULICS/Rijkswaterstaat. A World Wide Cost Estimate of Basic Coastal Protection Measures, ANNEX D in: *Strategies for Adaptation to Sea Level Rise,* Report of the Coastal Zone Management Subgroup, Intergovernmental Panel on Climate Change, Response Strategies Working Group, November, 1990 (Delft: 1990).

4. IPCC (1991). *The Seven Steps to the Assessment of the Vulnerability of Coastal Areas to Sea Level Rise — A Common Methodology,* Advisory Group on Assessing Vulnerability to Sea Level Rise and Coastal Zone Management, 20 September 1991, Revision no. 1 (The Hague: Rijkswaterstaat, Ministry of Transport and Public Works, 1990).

5. IPCC (1991). *Global Climate Change and the Rising Challenge of the Sea,* Response Strategies Working Group, Coastal Zone Management Subgroup (The Hague: Rijkswaterstaat, Ministry of Transport and Public Works, 1990).

6. Hoozemans, F. M. J., M. Marchand, and H. A. Pennekamp, *Sea Level Rise, A Global Vulnerability Assessment: Vulnerability Assessment for Population, Coastal Wetlands and Rice Production on a Global Scale,* Second Revised Edition (De Voorst/Delft: DELFT HYDRAULICS/Rijkswaterstaat, 1993).

7. Smith, K. *Environmental Hazards, Assessing Risk and Reducing Disaster* (London: Routledge, 1992).

8. *World Bank Annual Report* (London: Oxford University Press, 1990).

9. Bird, E. C. F. *Coastline Changes* (New York: John Wiley & Sons, 1985).

10. Fairbridge, R. W. *The Encyclopedia of Geomorphology,* Volume III (New York: Reinhold Book Corporation, 1968).

11. Hogben, N., N. M. Dacunha, and G. F. Olliver, *Global Wave Statistics* (Surrey: Unwin Brothers Limited, 1986).

12. Hurdle, D. P. and R. J. H. Stive, *Revision of SPM 1984 Wave Hindcast Model to Avoid Inconsistencies in Engineering Application* (Amsterdam: Elsevier, 1989).

13. Hoozemans, F. M. J. and J. Wiersma, Is the mean wave height in the North Sea increasing? *The Hydrographic Journal* 63, January 1992, (1992).

14. *Admiralty Tide Tables.* Volumes I, II, III (London: The hydrographer of the Navy).

15. Da Vinci. *Computation of Potentially Flooded Areas due to Sea Level Rise* (Chaumont-Gistoux, Belgium: 1992).

16. UNEP. *Global Resource Information Database* (Geneva: UNEP, 1991).

17. CERC. *Criteria for Evaluating Coastal Flood-Protection Structures,* Final Report, 89-15 (Washington, D.C.: CERC, 1989).

18. Weide, van der, J. and F. M. J. Hoozemans, *Tools and Techniques to Simulate Changes in Coastal Systems due to Climate Change* (De Voorst/Delft: DELFT HYDRAULICS, 1993).

19. The Economist. The World in Figures (London: 1987).

20. Encyclopedia Britannica, Inc. (1987, 1990). *Book of the Year 1987, 1990* (Chicago: Britannica World Data, Annual, 1987, 1990).

21. Times Books, *The Times Atlas of the World* (London: Times Inc., 1988).

22. World Almanac (1990). *The World Almanac and Book of Facts,* (New York: 1990).

23. *World Population Projection.* (John Hopkins University Press: 1989–1990).

24. Frassetto, R., *Impact of Sea Level Rise on Cities and Regions* (Venice: Centro Internazionale Citta d'Acqua, 1991).

25. Turner, R. K., P. M. Kelly, and R. C. Kay, *Cities at Risk* (London: BNA International INC).

26. U.S. Department of Commerce Bureau of the Census, World Population Profile: 1990, Washington, D.C. (1989).

CHAPTER 8

Impact of Climatic Change on Coastal Cities

Tjeerd Deelstra

THE PROBLEMS

Cities are both causes and victims of climatic change. Cities emit large volumes of greenhouse gases (Figure 8–1), coming from sources such as traffic, industry and households.[1] An example is the Netherlands, where research has shown that more than 80% of all greenhouse gases originate from cities (Table 8–1).[2] Climate change may manifest in:[3]

- Higher temperatures
- Rising sea levels
- Increased frequency and intensity of storms, rainfall, and floods
- Changes in ecosystems

HIGHER TEMPERATURES

Rising temperatures are a direct consequence of the growing concentration of a number of gases in the troposphere. In this lowest layer of the atmosphere, the part where weather phenomena occur, are various gases that disturb the earth's radiation balance. The greenhouse effect intensifies as the concentration of gas increases, of course. Extrapolation of currently available data predicts a situation in 2030 in which the various greenhouse gases will produce twice as much radiation as carbon dioxide did in the pre-industrial age. This results in higher temperatures in cities. Consequently, the higher temperatures have impacts on the built environment, people, and nature in cities, which differ for each type of climate, landscape, and season. By and

0-87371-301-X/95/$0.00+$.50
© 1995 by CRC Press, Inc.

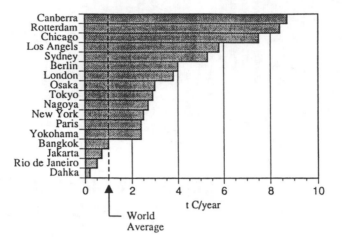

Figure 8–1 Per capita CO_2 emission of selected cities (ton C/year, 1985). (From Nishioka Shuzo, Yuchi Moriguchi, and Sombo Yamamura. In: James McCulloch, Ed. *Cities and Global Change*. Climate Institute, Washington. 1991. With permission.)

large, warmer types of climate are less prone to rising temperature in winter than colder types, and vice versa in the summer. In the future, desert areas in Africa, for instance, will have a higher constant temperature, while areas such as central and northern Europe will be considerably warmer in winter but less so in summer. Differences in winter temperatures will be relatively smaller; differences in summer temperatures will be larger.

Higher temperatures due to climate change will alter the water balance of cities, resulting in water logging or shortages. In association with rising temperatures, river levels will sink because of disappearing glaciers that feed rivers and drier soil. There will be fewer clouds, from which less rain will fall; solar-radiation will encourage evaporation. The cases of the Hudson and Rhine Rivers form examples. It has been calculated[4] that a water deficiency of 28 to 42% of the planned supply in the Hudson River will occur when CO_2 levels are doubled. The lowest summer flow of the Rhine might sink by 20% to 500 m³ s.[5]

The microclimate of bigger cities is already different from rural areas (Table 8-2).[6] Cities are warmer, have more dust and pollution in the air above the city, and are drier. The "urban heat island" causes heat stress from which older people in particular suffer. The hotter the city, the more these problems will increase. It has been calculated that a 4°C temperature increase in the San Francisco Bay area would increase maximum ozone concentrations by 20% and the number of exposure hours for people would triple.[7] A rise in temperature also leads to smog, which has immediate impacts on people's well being, causing lung abnormalities in the long term.[8] Research has shown that ozone stimulates lung tumors.[9] In warmer cities new types of vector-based diseases such as malaria could occur. When the need for cooling in hotter cities is met by using fossil fueled installations, the air above cities will be polluted with more photochemical oxidants and aerosols, changing visibility

Table 8–1 Sources of Emissions of Greenhouse Gases in the Netherlands

Gas	Source	Neth.	Urban	Rural	
CO_2	Industry and manufacturing	19	19	—	
	Traffic and transport	14	10	4	(1)
	The built environment	13	12	1	(2)
	Energy supply and distribution	29	29	—	
	Others (e.g., waste incineration)	10	10	—	
	Bioorganic sources	15	—	15	
	Total	100	80	20	
CFCs	Industry/consumers	100	90	10	(2)
CH_4	Extraction and transport				
	of oil and gas	16	—	16	
	Waste dumps	27	24	3	(2)
	Natural marshes and waters	15	—	15	
	Animal waste/manure	7	—	7	
	Ruminating animals	35	—	35	
	Total	100	24	76	
N_2O	Combustion of fossil fuels	16	14	2	(2)
	Water purification installations	8	8	—	
	Grass lands	60	—	60	
	Coastal and open waters	16	—	16	
	Total	100	22	78	
NO_x	Traffic and transport	60	42	18	(1)
	Energy and industry	30	30	—	
	Others	10	9	1	(2)
	Total	100	81	19	
CO	Combustion engines (traffic)	70	49	21	(1)
	Combustion engines (other)	25	25	—	
	Other	5	4	1	(2)
	Total	100	78	22	
VOC	Partial combustion (traffic)	40	28	12	(1)
	Industry, small companies and				
	households	45	41	4	(2)
	Other	15	14	1	(2)
	Total	100	83	17	

From Deelstra, Tjeerd and Robuit de Bosch. *The Greenhouse Effect. Preventative Urban Activities in the Netherlands.* The International Institute for the Urban Environment, Delft. 1993. With permission.

and albedo. This and increased acid rain will cause the quality of life to deteriorate. Urban trees and vegetation will also be affected.

RISING SEA LEVELS

Many cities have developed along coasts. It is estimated that about 80% of cities will be exposed to risk from rising sea levels.[10] Consequences might be: inundation, silting, beach erosion, coastal damage (and damage to related industries and harbors), flooding and saltwater intrusion, land loss and habitat modification, and damage to road infrastructures, drainage systems, and buildings. An example is Bangkok, where groundwater extraction is presently causing subsidence at levels that are among the highest in the world.[11] With the rising of sea levels the city will be in danger of continuous flooding. Alexandria is another example. The city already suffers problems of subsidence due to groundwater extraction, natural beach erosion, and sediment

Table 8–2 Comparison of Climate Parameters of Cities and Surrounding Areas

Climate parameters	Characteristics	In comparison to the surrounding area
Air pollution	Condensation	10 times more
	Gaseous pollution	5-15 times more
Solar radiation	Global solar radiation	15-20% less
	Ultraviolet radiation, winter	30% less
	Ultraviolet radiation, summer	5% less
	Duration of sunshine	5-15%
Air temperature	Annual mean average	0.5-1.5°C
	On clear days	2-6°C higher
Wind speed	Annual mean average	10-20% less
	Calm winds	5-20% more
Relative humidity	Winter	2% less
	Summer	8-10% less
Clouds	Overcast	5-10% more
	Fog (winter)	100% more
	Fog (summer)	30% more
Precipitation	Total rainfall	5-10% more
	Less than 5 mm rainfall	10% more
	Daily snowfall	5% less

From Landsberg, Helmut E. *The Urban Climate.* Academic Press, New York. 1991. With permission.

starvation as a result of the Aswan High Dam. The city could possibly become a peninsula due to further subsidence.

Some low-lying countries, like Bangladesh and the Netherlands, have learned to live with risks. But only the richer countries will be able to cope with rising sea levels. The costs of implementing preventative measures for harbors and cities in the Netherlands (the majority of which is protected by dikes) are calculated at U.S. $40.9 billion. This estimate does not include water management costs or costs incurred in the maintenance of groundwater levels.[13]

INCREASED FREQUENCY AND INTENSITY OF STORMS, RAINFALL, AND FLOODS

The predicted effects of climate change include increased frequency of storms, rains, and floods and a generally less stable local/regional climate. Avalanches, cyclones, earthquakes, landslides, typhoons, hurricanes and storms, cold waves, and heat waves are expected to increase. It has been analyzed that storms in the San Francisco Bay now numbering one per century will increase to one per decade.[14]

Floods can cause serious damage. Take the example of cities in the Ganges Delta. Calculations show that increased floods as a result of climate change could result in loss of a quarter of the land that is currently inhabited. Research funded by the World Bank and the United Nations Development Programme attempts to strengthen traditional methods of flood hazard adjustment, which allow shallow flooding.[15]

In Hamburg, water defense relies on a series of earthen and concrete embankments around the low-lying parts of the city (where approximately 320,000 people

Table 8–3 Costs for Protecting Cities to
Impacts of Climate Change

Region	Cost (billion U.S. $)	
	Cities	Harbors
N. America	25.3	5.4
C. America	0.1	0.1
Caribbean Islands	1.3	2.2
S. America (Atlantic)	5.7	2.6
S. America (Pacific)	1.4	0.2
Atlantic Ocean (small isl.)	0.1	0.0
N. and W. Europe	12.5	5.5
Baltic Sea Coast	3.0	1.1
N. Mediterranean	6.5	2.2
S. Mediterranean	3.0	1.1
Africa (Atlantic)	3.7	1.2
Africa (Indian Oc.)	1.9	0.9
Gulf States	0.8	3.5
Asia (Indian Oc.)	8.2	0.7
Indian Oc. isl.	0.3	0.0
Southeast Asia	4.8	1.5
East Asia	14.9	7.9
Pacific Oc. (large isl.)	6.2	1.8
Pacific Oc. (small isl.)	0.2	0.1
USSR	1.8	1.0
Total	102.6	40.9

From Hootsman, M. J. Delft Hydraulics Report
H1068. 1990. With permission.

live). Plans are now being made for the construction of a 300-m wide barrier across the river Elbe to protect the city from increased risks of flooding.[16]

Estimated costs of such projects differ greatly, but total amounts needed to protect the world's major cities from danger of flooding can be estimated in the region of U.S. $102.6 billion (Table 8–3).[17]

CHANGES IN ECOSYSTEMS

A rise in sea level could increase landward penetration of marine and maritime influences. "Wash lands" could be lost, reducing the space available to birds and other animals. Salt penetration could cause fish depletion and longer stretches of rivers could become brackish. This would affect food chains and consequently, bird populations. Earthen cliffs could erode resulting in loss of habitats.[18]

Some of the natural systems are vital to towns and cities, which harvest food from them. It is expected that human settlements in Mauretania, Namibia, Peru, and Somalia will be seriously affected by the inevitable collapse of the fishing industry due to climate change.[19]

Parks and trees in cities will also be affected. It is believed that climate change will occur at such a rate that natural adaptation and migration abilities of plants and trees may not be able to keep pace. It will therefore take considerable costs to replant urban vegetation appropriate to hotter cities and towns, in particular in cities where evaporation will also increase and groundwater tables fall.

THE CASE OF EUROPE

The general effects of rising temperature will be the same everywhere: sea level rise and greater aridity on land. According to the General Circulation Model (GCM) developed by the Goddard Institute of Space Sciences (GISS) in New York, sea level will rise by 25 to 165 cm due to expansion and to melting land-ice. Rapidly melting polar ice, which becomes lighter as a result, will cause the land to rise in the north of Europe, Asia, and America, and also possibly in the southernmost parts of South America, Australia, and New Zealand. This isostatic elevation is expected to be 10 to 20 cm per century, and will partly compensate for the rising sea level. Further toward the equator, however, land will subside.

The interaction of higher temperatures and water on and around the earth is so complex that the precise consequences will have to be examined per individual location. A cautious and speculative move is made in this direction in this section, which discusses impacts on selected coastal cities in and near Europe. This section draws upon a study, executed by the International Institute for the Urban Environment for the Netherlands Ministry of Housing, Planning and Environment.[20] For the sake of convenience an average temperature increase of 3.5°C and an average rise in sea level of 1.00 m are envisaged. For several cities calculations were made for precipitation and temperature, based on Walther and Lieth's World Climate Atlas of 1967 and the GISS-GSM. Expected impacts of a doubling of CO_2 levels are described here for the cities of Amsterdam, Archangel, Barcelona, Gothenburg, Istanbul, Liverpool, Saint Petersburg, and Venice (Figure 8–2).

AMSTERDAM (NETHERLANDS)

Amsterdam, the Dutch capital, is situated in the west of the Netherlands where the river Amstel enters the IJsselmeer. Amsterdam's history goes back to 1275, when a penetrable dam was built in the mouth of the river Amstel, an event which gave the oldest part of the town its name. The settlement spread along the banks of the Amstel and was soon intersected by the canals that formed its early fortifications. At first the street pattern spread out fairly symmetrically on either side of the river. Later the town expanded in a more southeasterly direction, its elongated shape evolving into a near-semicircle. In the 17th and 18th centuries rings of canals, based on a 1665 development plan, were built. In the 19th century Amsterdam grew like a shell from this concentric arrangement, expanding radically in the 20th century to become a large urban area. Its chief economic significance, apart from industry and commerce, lies in the service sector. The river Amstel still flows into the IJsselmeer, which is now dammed off from the North Sea by sluices. New canals for water transport connect Amsterdam with the North Sea and the Atlantic. The urban agglomeration now counts about one million inhabitants.

The landscape of the Netherlands is part of the North German Plain. Some 30% of the country is below sea level. The soil around Amsterdam consists of strata of peat, clay, and sand, load-bearing sand strata being at depths of approximately 11 and

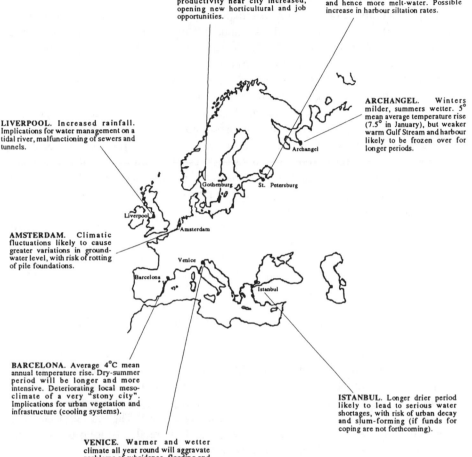

GOTHENBURG. Milder winters and wetter summers. Implications for investments in municipal heating systems (less need) and in pumping and drainage (greater need). Agricultural productivity near city increased, opening new horticultural and job opportunities.

ST. PETERSBURG. Increased precipitation and milder winters, with more snow-fall and hence more melt-water. Possible increase in harbour siltation rates.

ARCHANGEL. Winters milder, summers wetter. 5° mean average temperature rise (7.5° in January), but weaker warm Gulf Stream and harbour likely to be frozen over for longer periods.

LIVERPOOL. Increased rainfall. Implications for water management on a tidal river, malfunctioning of sewers and tunnels.

AMSTERDAM. Climatic fluctuations likely to cause greater variations in ground-water level, with risk of rotting of pile foundations.

BARCELONA. Average 4°C mean annual temperature rise. Dry-summer period will be longer and more intensive. Deteriorating local meso-climate of a very "stony city". Implications for urban vegetation and infrastructure (cooling systems).

ISTANBUL. Longer drier period likely to lead to serious water shortages, with risk of urban decay and slum-forming (if funds for coping are not forthcoming).

VENICE. Warmer and wetter climate all year round will aggravate problems of subsidence, flooding and groundwater levels.

Figure 8–2 Location of selected cities in Europe of which impacts of climatic change are described. (From Ecology Chronicle UNESCO. 1991.)

18 to 19 m below "New" or "Exact Amsterdam level", the ordinance datum from which all height is measured in the Netherlands. Traditional building foundations are wooden piles, which usually rest on one of the two layers of sand. Construction is light and airy. Because of foundations in the old town and in a large amount of prewar construction, the groundwater level is very important for Amsterdam.

The Dutch climate is moderate and wet, without a definite dry period. The city is situated in a marine climate region. The climate near Amsterdam is influenced by both the North Sea and the IJsselmeer, the dammed-off, freshwater inland lake. The average temperature is 2°C in January, 18.5°C in July, and 10°C mean annual. Precipitation for these periods is 52, 64, and 648 mm. March is driest, and the coast is wettest in autumn.

Expected climatic change in Amsterdam involves an average rise in temperature of 5°C in January, 3°C in July, and 4°C annually. Average precipitation for the same periods is expected to increase by 16, 8, and 144 mm. Winters, especially, will become warmer and wetter, summers less so.

Climatic fluctuations are — among others — expected to cause greater variations in groundwater level: higher in winter and lower in summer. This means that basements and cellars in Amsterdam are liable to flood in winter, increasing damp- ness and the chance of fungus. This will affect people's health and the substance of their houses. Things will be worse in the warm months, when the water sinks below the tops of the — mainly wooden — piles. At present the piles are constantly under water, a conservation method that threatens to fail in future summers. In due course the piles will rot and gradually affect foundations, causing subsidence and cracks and eventually necessitating demolition.

Furthermore, the problem will not be confined to Amsterdam but will affect all low-lying historical towns in the western Netherlands, Belgium (Antwerp, for in- stance), and comparable locations elsewhere. It is technically possible to cope with these problems by changing the system of water management but this will require a lot of money, planning, and organization.

ARCHANGEL (COMMONWEALTH OF INDEPENDENT STATES, C.I.S.)

Archangel lies in the north of the C.I.S. where the North Dvina enters the White Sea. The population is about half a million. The geographical landscape around Archangel is part of the East European Plain, a flat region that rises slightly towards the east, where it is bounded by the Urals. The average height of the plain is 200 m, but Archangel was built only a few meters above sea level. The harbor, accessible to sea-going vessels, is frozen half the year round, its relatively long open period being due to the warm gulf stream that penetrates the White Sea.

The climate around Archangel is characterized by warm summers and cold winters, without a definite dry period. The city lies on the border of a continental and a polar climate zone. The average temperature is –12.5°C in January, 14°C in June, and +0.4°C mean annual. Average precipitation for the same periods is 21, 65, and 466 mm.

Expected climatic change in Archangel involves an average rise in temperature of 7.5°C in January, 3°C in July, and 5°C mean annual. Precipitation in the same periods will increase by 16, 32, and 288 mm. Winters will be milder and summers wetter.

As a result of climatic change, the warm gulf stream towards and in the White Sea will probably become weaker. The harbor is likely to be frozen over for longer periods. More snow will fall in Archangel and its environs, and remain longer on the ground. When it melts in late March/early April, peak runoff will increase. Dust particles in the snow, sand, and other eroded rock components and the like will be deposited in the river and harbors, which will become shallower. The effect could be heightened by land-rise as a consequence of the dwindling ice cap. Another consequence of the increased quantity of snow is that the harbors will gradually become less saline, making them more liable to freeze over.

All in all there is a chance that Archangel will lose its function as an important port. Wetter summers could affect the quality of life. Lakes and pools will form in the impermeable ground, providing breeding places for insects, mosquitoes, and suchlike. Besides being unpleasant, these are health hazards, as has been experienced in Siberia where certain insects, ticks, and so forth have affected public health in damp periods.

Some of the problems could be solved by adopting technical measures (dredging harbors, sending icebreakers into shipping lanes, draining lakes and pools, etc.). This calls for considerable investment. A more obvious solution would be to relocate Archangel's present activities and to develop new ones in the existing city, such as local industries and a new technology of urban surface water management combined with running off the greater amount of summer precipitation.

BARCELONA (SPAIN)

Barcelona is situated in northeastern Spain, on the Mediterranean coast. It is the capital of the province of the same name, and also of the region of Catalonia. Catalonia is a mountainous and fairly infertile landscape — with the exception of an area containing several rivers — bounded by the Pyrenees and the Ebro Estuary. East of the Catalonian coastal range, Barcelona lies in a flat coastal area north of the Llobregat Delta. This coastal area is surrounded by granite and slate hills, including the Tibidada (532 m), and is part of an irrigation region bordering the Mediterranean Sea.

Barcelona formed at the foot of the limestone hill Montjuich (191 m), upon which a fortress was later built. Until 1858 the medieval center was surrounded by a hexagonal wall. After the fortifications were pulled down, boulevards were built around the historical center, and the city was expanded, using a grid of 133 × 133 m squares with streets 20 m wide. There is not much green space in the city. Barcelona is Spain's second most important port and industrial city. The harbors are artificial, and are located in the suburb of Barceloneta. Barcelona's population is now nearly two million.

The climate around Barcelona is moderate and wet, with a dry period in summer. A little north and west of the city this subtropical climate gives way to a marine type. The entire east coast has hot summers with a lot of sun. The average temperature is 9°C in January, 24°C in July, and 16°C annually. Average precipitation for the same periods is 29, 24, and 600 mm. The dry period notwithstanding, most rain falls in late summer.

Expected climatic change in Barcelona involves an average rise in temperature of 4°C in January, July, and annually. Precipitation in the same periods will increase by 8, 0, and 48 mm. The dry summer period will be longer and more intensive.

Expected climatic changes may cause serious problems connected with the deteriorating local climate. The original idea of the structure of the 19th century development plan was to realize a grid of squares with cut-off corners. Buildings were planned on only two sides of the blocks making the inside areas that connect with the streets into gardens. This concept was abandoned under the pressure of rising land prices. Some blocks consist entirely of buildings, the only vegetation being the trees lining the roads. This makes Barcelona a very stony city in which even now temperatures soar in the hot, dry summers. It is already a heat-island with much higher temperatures than the surrounding countryside. This is a direct consequence of the mass of built-up area, coupled with a lack of cooling and ventilation in the city. There are two reasons for these deficiencies: the city does not have enough vegetation and there is not enough surface water, two climate regulators which would improve the air's humidity, circulation, and oxygen content.

If temperatures rise further as a result of expected climatic change, streets and buildings will store even more heat, practically eliminating nocturnal cooling. Drinking and cooling water will be in short supply, due to summers that will get longer and much drier. There will not be enough to water the sparse urban vegetation. Sewers will dry up and smell, and there will be an increased danger of disease. Summers in Barcelona will be neither pleasant nor healthy for the city's inhabitants and tourists. More heat waves are likely, with fatal results for people unable to stand the heat.

To cope with the impacts of climatic change, desert-hardy plants could be introduced, buildings could be air-conditioned, and compressed air could be forced into sewers. All this calls for investments on what is probably an unfeasible scale: the city's entire infrastructure is involved. Air-conditioning systems, which are expensive and vulnerable installations, currently require fossil fuel. During power cuts the entire city is helpless. New systems would have to be devised (sea water cooling, for example). The city might also wish to build "wind towers" to harness every breath of wind to cool the air.

GOTHENBURG (SWEDEN)

Gothenburg lies in southwestern Sweden on the Göta Älv Estuary in the Kattegat. It is the capital of the Göteborg och Bohus district. In 1985 the city's population was 425,495 and that of the urban agglomeration 704,052.

The landscape of southwestern Sweden contains the gneiss and granite plateaus of Täberg (343 m), gradually rising towards the southwest and southeast in Småland and Blekinge. The coastal regions are the most fertile and, hence, most densely populated. The interior is less accommodating, due to rock breaking through to the surface. Gothenburg's urban topography is also marked by areas of bare rock and lower-lying parts with an unstable clay layer. Off the coast of Göteborg och Bohus are a large number of islands (the Skargárd archipelago). The climate around Gothenburg is moderate and wet, without a definite dry period (Köppen climate system Cf). Just north and northwest of the city this marine climate gives way to continental climate. Gothenburg harbor, the most important in Sweden, is ice-free almost all year round.

Gothenburg is a planned city. It was built for Charles IX and destroyed by the Danes in 1607. In 1619 Gustaf Adolf II decided to rebuild Gothenburg in a new location on the Göta Älv. The new plan envisaged regular blocks of buildings, a system of canals and sturdy, bastioned fortifications. The city became an important trading and shipping port and is also a major industrial center today. It has expanded on both sides of the river since the turn of the century, with large housing estates built in the 1960s.

Gothenburg's buildings are heated by an ingenious municipal system fed by energy from the central garbage incineration installation. Additional heat is obtained from wastewater from the sewers and residual heat from the oil refineries in the port.

The average temperature is +0.5°C in January, 16.5°C in July, and 7.7°C annually. Average precipitation for the same periods is 57, 71, and 738 mm. The ground is snow-covered for about a month.

Expected climatic change in Gothenburg involves an average rise in temperature of 5.3°C in January, 2.5°C in July, and 4°C annually. Precipitation in the same periods will increase by 16, 16, and 192 mm. Winters will become milder and summers wetter.

Because of these changes in climate, investment in the municipal heating system will cease to be worthwhile, since buildings will require less heat. This means reorganizing a system that has admirably proved its efficiency. Mild winters and wet summers could result in increased agricultural productivity in and near the city, creating jobs and even export possibilities. Certain crops that cannot be cultivated at present will become available to the people of Gothenburg, with possible effects on local economy and habits.

Increased precipitation, combined with a possible rise in sea level, may also affect drainage. Natural drainage via canals will be hampered. Extra canals and pumping stations could provide a solution. This, naturally, requires careful planning and financing.

ISTANBUL (TURKEY)

Istanbul is in northern Turkey and is built on both sides of the Bosporus, the passage between the Black Sea and the Mediterranean. It is the capital of the province of the same name and has 5,475,982 inhabitants (1985).

The landscape around Istanbul is part of the foothills of the Istranca Daglari. It is rolling country, and Istanbul, like Rome, was built on seven hills. The Mediterranean region in which Istanbul is situated is tectonically active.

The old center of Istanbul was built between 324 and 330 A.D. on the site of the former Byzantium. It was named Constantinople after its founder, Constantine the Great, who made it capital of the Roman Empire. In 412 Constantinople started to expand westward, and was enclosed by the walls which still stand today. In 537 the Hagia Sophia was consecrated. Captured by the Turks in 1453, the city became part of the Ottoman Empire. The Tokapi Palace and the Blue Mosque (1610) date from this period. Constantinople ceased to be the capital of Turkey in 1923, when its name was officially changed to Istanbul. Besides a large number of palaces and about 200 mosques, there is a maze of narrow, winding streets.

The climate round Istanbul is moderate and wet, with dry summers (Köppen climate system Cs). It is a subtropical climate, becoming marine toward the north-west of the city. The average temperature is 5.5°C in January, 22°C in July, and 13.6°C annually. Average precipitation for the same periods is 79, 22, and 633 mm.

Expected climatic change in Istanbul involves an average rise in temperature of 4°C in January, July, as well as annually. Precipitation in the same periods will remain the same. The dry period will be longer.

Istanbul is a European city in a largely Asiatic country. This makes it a meeting place for different cultures and an easy and obvious place from which to step into the rich western world. The expected climatic changes will cause the economic situation in Turkey to deteriorate; harvests will run into problems and deserts will form. Country dwellers unable to subsist in the agrarian sector will very likely seek work in the cities. People will also tend to migrate further north to the richer areas of Europe, where the changing climate will not have such a grave effect on survival chances. The migration movement could converge on Istanbul. People are likely to move from rural areas to European Istanbul, whose inhabitants in turn will become increasingly oriented toward other European countries. There will be a shift in urban population, coupled with diminishing prosperity. This will create a self-sustaining downward spiral. Impoverishment will cause facilities and, hence, life in the city to get seriously out of hand and could eventually ruin the city.

Measures aimed at reversing the process outlined here call for tremendous investments and a highly determined and extremely well-organized public adminis-tration.

LIVERPOOL (U.K.)

Liverpool lies in western England, in the county of Lancashire. It is situated on the Mersey Estuary in the Irish Sea and has a population of 491,500 (1985). The landscape around Liverpool belongs to the Midlands, which in turn are part of the British Lowlands. The Mersey, like other navigable rivers, is deep-bedded, has little fall and is hence highly suitable for navigation. While estuarial silting-up is usually compensated by tidal erosion, the Mersey is affected by a large tidal fluctuation of 10 m and heavy silt deposits.

Liverpool grew rapidly during the British industrial revolution. The characteristic docks of that period have been restored as monuments of industrial archaeology. High buildings are other typical features of the industrial revolution, which brought great wealth to this second most important city of the British Empire. Road and rail tunnels under the Mersey connect the districts that developed on both banks in the course of the 19th and 20th centuries.

The climate around Liverpool is moderate and wet, without a definite dry period (Köppen climate system Cf). The city lies in a marine climate zone. The average temperature is 4.5°C in January, 15°C in July, and 9.6°C annually. Average precipi-tation for the same periods is 63, 66, and 737 mm. October has the heaviest rainfall, March the least.

Expected climatic change in Liverpool involves an average rise in temperature of 4°C in January, 3.5°C in July, and 4°C annually. Precipitation in the same periods will increase by 8, 0, and 48 mm.

Possible impacts of climatic change here will primarily manifest in water management. Increased precipitation may raise water level in the docks, making them inaccessible at times and putting paid to their function as a monument and museum. The Mersey tunnels are liable to be affected by increasing groundwater pressure. This problem is already apparent, and is due to the disappearance of many old industries that used to pump out the groundwater. Water is already oozing into the tunnels. If pressure increases there will be cracks and flooding.

Another problem of rising water levels affects sewers. The municipal sewers empty into the Mersey, and the dirty water is carried off to sea at ebb tide. When levels rise the tidal pattern will change and many sewers will no longer be emptied by tidal reflux.

Possible solutions would be to close off the docks with a dam, which would have its aesthetic drawbacks, to replace the old tunnels with new ones, and radically change the sewer system (or to pump the sewers dry by artificial means, in combination with other technical measures).

All this costs a lot of money which will be hard to find in Liverpool. The disappearance of obsolete industries has led to high unemployment. Liverpool is not the rich city it once was.

SAINT PETERSBURG (COMMONWEALTH OF INDEPENDENT STATES C.I.S.)

Saint Petersburg lies in the western part of the C.I.S. on the Neva estuary in the Gulf of Finland. It is the capital of the province of *Oblast* with the name Leningrad. In 1985 it had a population of approximately 4,567,000.

The city of Saint Petersburg was founded in 1703 by Peter the Great. It was built where the river Neva splits up into (from north to south) the Great Neva, the Little Neva, and the Great Nevka, which divide the city into four separate parts. Development began on the Great Side, south of the Great Neva, with three radial main roads, intersected by a system of canals. Buildings in the former royal capital adhere to a variety of classicist styles.

Today Saint Petersburg is the biggest shipping port in the C.I.S., a major industrial city, and the country's second most important cultural and scientific center. There are a large number of parks and gardens. In the 1960s and 1970s Saint Petersburg was enlarged by the addition of a reclaimed stretch of coast about 25 km long and 1 km wide.

The landscape around Saint Petersburg is part of the East European Plain, a fairly flat region that rises slightly towards the east, where it is bordered by the Urals. The average height of this plain is 200 m, but Saint Petersburg itself is only 1 to 2 m above sea level. The city spreads over a total of 101 islands in the Neva Delta in a geographically unfavorable region prone to frequent flooding. The soil is not very solid, so that buildings must rest on piles.

The climate around Saint Petersburg is characterized by warm summers and cold winters, without a definite dry period (Köppen climate system Df). The city is on the border of a continental and mountain climate region. The average temperature is 7°C in January, 17.5°C in July, and 3.7°C annually. Average precipitation over the same periods is 24, 64, and 470 mm. Most rain falls in July, March being the driest month.

Expected climatic change in Saint Petersburg involves an average rise in temperature of 6°C in January, 2°C in July, and 4°C mean annual. Precipitation in the same periods will increase by 16, 32, and 288 mm. Winters, especially, will be warmer, and summers much wetter.

Since the end of the last Ice Age, land has been rising. In Saint Petersburg, as in all northern cities of Europe, land is expected to rise as a result of climate change. For a low-lying, flood-prone city like Saint Petersburg, this is a favorable circumstance, since sea level rise will have drastic effects.

Increased precipitation and milder winters will lead to more snowfall and, hence, more melt water, which the rivers will have to carry off. Snow retains a relatively large amount of erosion components that are likely to be deposited in the estuary, where water flows more slowly. Saint Petersburg's harbors are therefore liable to become inaccessible to ships in times of drought.

Increased snowfall can also deposit more dust particles from the already polluted air onto streets and buildings, with dire consequences. The city's vegetation is also at risk. If more snow falls, more salt will be strewn, harming vegetation and trees in the streets. Parks and fields will turn into expanses of mud. Feasible solutions such as harbor excavation, electric road-heating, and other types of street vegetation are expensive and call for a great deal of planning and organization.

VENICE (ITALY)

Venice is in North Italy on the Adriatic Sea. It is the capital of the province of the same name and of the region of Veneto. Including the mainland industrial district of Mestre, the city has 339,272 inhabitants (1984), about 100,000 of whom live in the old town.

Geographically, Veneto is part of the Po Plain, an approximately 500-km long area that widens toward the east. It is a very fertile region that is watered by the Po, Italy's most important river, and its tributaries. Flooding is frequent. Venice is built on a total of 118 islands, which are linked by about 400 bridges. The islands lie in the lagoon, which is separated from the Adriatic by land-fingers (lidi).

The climate around Venice is moderate and wet, without a definite dry period (Köppen climate system Cf). The city lies in a marine climate zone, although the Po Plain has hot summers and cold, but wet winters, with a lot of fog and mist. The average temperature is 3°C in January, 22.5°C in July, and 13.4°C annually. Average precipitation for the same periods is 43, 58, and 757 mm. July is the wettest month.

Venice's gradual development dates from the 6th century. At first it consisted of a dozen rival lagoon towns. It took until 779 for the first doge to be elected and the city-state to become a fact. The mercantile city's flourishing economy dates from the Middle Ages, when the Crusades opened up new markets.

Due to Venice's situation on a conglomeration of small islands, most traffic is waterborne. The city's structure is determined by the Canal Grande, almost 4 km long and an average 70 m wide, which winds its way from the southeast to the northwest — from the Doge's Palace on St. Mark's Square to the railway station. In 1846 Venice was linked to the mainland by a railway bridge, and in 1933 by a road bridge. The historical center has a mainly touristic and cultural function. Most Venetians and jobs are to be found in Mestre and Maghera. As recently as 1966 Venice was almost completely inundated by a storm flood. Subsidence and water and air pollution are threatening the city's survival.

For a city like Venice, which is not only surrounded by water but also intersected by numerous canals, subsidence is a major problem. Although some attribute subsidence to movements in the subterranean stone strata, and others to the extraction of natural gas in the Po Delta, most experts agree that the prime causes are the pumping-up of groundwater under the city and the construction of large industrial complexes on the mainland. A connection has already been established between the degree of subsidence and the quantities of water drawn from roughly 7000 wells.

Expected climatic change in Venice involves an average rise in temperature of 4°C in January, 3.5°C in July, and 4°C annually. Precipitation in the same periods will increase by 8, 8, and 96 mm. It will be warmer and wetter all year round.

The expected climatic changes will aggravate the aforementioned problems of subsidence. Higher temperatures and faster evaporation will increase water consumption, causing the groundwater level to sink even more rapidly. Besides the additional subsidence resulting from this, the city faces a gradual rise in sea level. The already very real danger of flooding will thus be considerably augmented and could lead to precarious situations not only for the population, but also for the survival of the city itself.

There would seem to be no acceptable solutions. The only expedient measure is to develop an extremely large-scale regional water management system (enabling the lagoon to control its own level). This involves serious financial, organizational, and aesthetic problems.

COASTAL CITIES AND THEIR LINKAGES WITH HINTERLANDS

Coastal cities cannot be observed in isolation. Cities are linked with each other through economic ties. Changes in coastal cities will affect their hinterlands, while for example impacts of climate change in mountainous inland areas will have its effects on cities along coasts.

Chain reactions will occur, in particular along rivers that connect coasts and inland areas. This can be illustrated with the example of the northwestern European megapolis, situated in the Rhine-Maas Basin (Figure 8–3).[21]

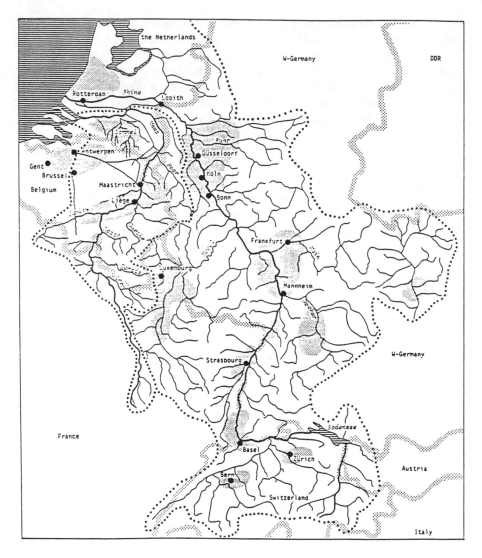

Figure 8–3 Urban concentrations in the Rhine-Maas basin and influenced land. (From Deelstra, Tjeerd, Johan Galjaard, and Rob H. G. Jongman. *Possible Impacts of Climatic Change on European Cities.* The International Institute for the Urban Environment, Delft. 1989. With permission.)

THE RHINE-MAAS BASIN

The northwestern European megapolis is densely populated. It has important political centers (Berne, Bonn, Luxemburg, Strasbourg, Brussels, The Hague), as well as commercial centers (Rotterdam, Frankfurt, Zurich), the world's greatest concentration of oil refineries (Rotterdam, the Ruhr area, Mannheim) and chemical plants (Basle, Strasbourg, Frankfurt-Mainz, the Rhine-Ruhr area, Rotterdam). Steel used to be a major industry (Metz-Nancy, Saarbrücken, Charleroi, the Rhine-Ruhr area). Though still relatively important there, other industries, notably chemicals and microelectronics, are growing.

Coal mining is on the wane. In Belgium, France, and North-Rhine Westphalia, older mines are closing down, as are many older steel and metallurgical plants. In North Rhine-Westphalia (the Rhine-Ruhr area), as elsewhere, this has led to high unemployment, but has also boosted possibilities of improving the quality of the urban environment. Such improvement is vital if the old industrial areas are to compete with more attractive locations such as Munich and Stuttgart.

In the Rhine-Ruhr area and Randstad Holland, plans for urban reconstruction and improvement are currently being implemented. Renewal of the urban fabric is an activity being performed in the cities of these areas. Abandoned harbors are being redeveloped for offices, houses, and cultural purposes.

In the Rhine-Ruhr area, recultivation of old lignitic open-face mines is taking place, mostly in the form of forests for recreational and ecological purposes. Recently abandoned coal mines and industrial plants are scheduled for similar restructuralization. A topical issue in the Ruhr area is "Renaturing", or restoration of nature to the river Emscher, at present a sewer.

Randstad Holland is currently the scene of "Randstad Groenstructuur", a structure of forests and woodland corridors in urban fringes with a mainly ecological and recreational function. Environmental improvement schemes for Charleroi-Liège have recently been announced.

Communication and transport in the Rhine-Maas Basin are well developed. The Rhine and Maas are important European shipping routes. In 1984 171 × 10^6 t of freight crossed the Dutch-German border on the Rhine. Freight on the Maas amounted to 2.3 × 10^6 t that same year.

Other forms of transport make use of a well-developed network of motorways, railroads, and pipelines. Major international and transcontinental airports are Frankfurt, Düsseldorf, Amsterdam, Zurich, and Brussels. Smaller airports are Cologne, Luxemburg, Rotterdam, Stuttgart, Strasbourg, Nuremberg, and Basle.

THE CLIMATE

The Swiss and Austrian climate is of Alpine origin, with annual precipitation ranging from 1800 to 2500 mm. In the German and French areas of the catchment, the climate is sub-Atlantic/subcontinental, manifest in the substantial differences in temperature and precipitation surplus.

There are many differences between rain-shadow areas (Colmar, Mainz) and high-precipitation areas (Black Forest). Precipitation in the more Atlantic climate of Belgium and the Netherlands is roughly the same as the German and French catchment areas, but temperatures are lower and evaporation is less. This produces an effective precipitation increase.

In the Alps in winter (December to February), most precipitation is stored as snow. Snow is also relatively important for water storage in central Europe, though not in the Atlantic lowlands.

Table 8–4 Discharge of the Maas between 1911 and
1960, Projected for $2 \times CO_2$ in the
Atmosphere, at Borgharen (Dutch-Belgian
Border) in m^3/sec

	1911-1960	$2 \times CO_2$
Mean summer discharge	130	100
Minimum summer discharge	0	0
Maximum summer discharge	1150	850
Mean winter discharge	390	470
Minimum winter discharge	0	50
Maximum winter discharge	2800	3300

From Deelstra, Tjeerd, Johan Galjaard, and Rob H. G.
Jongman. *Possible Impacts of Climatic Change on European Cities.* The International Institute for the Urban Environment, Delft. 1989. With permission.

THE LANDSCAPE AND ITS HYDROLOGY

The Maas catchment is 33,000 km^2. There is not much snowfall in winter, which
is why the Maas is known as a rain river. The Maas rises in northeastern France.
Maximum discharge is in January; minimum discharge is between July and September. The discharge pattern is subject to great interannual variation, though. Soil in the
Ardennes is fairly impermeable, causing rapid and superficial runoff of much of the
water after rain. This is the main reason for the fast decrease in river flow during dry
periods (Table 8–4). Canals supply the Brussels-Antwerp region and some southern
parts of the Netherlands with Maas water.

The Rhine catchment is 185,000 km^2. The largest part is in western Germany:
160,000 km^2; in France and Switzerland the catchment is 20,000 and 5000 km^2,
respectively. In the Netherlands canals serve 25,000 km^2 directly or indirectly with
Rhine water. Other regions in the catchment are Luxemburg, Vorarlberg (Austria),
Liechtenstein, and parts of Germany, Italy, and Belgium.

Most of the upper and middle reaches of the catchment have impermeable soil
(Alps, Black Forest, Jura, Sauerland). The central valley (Basle to Mainz) and the
entire lower area downstream of Bonn have more permeable soil, with large aquifers
and water tables close to the surface. Upstream (Alps to the Black Forest) summer
discharge is high and winter discharge low. Downstream of the Black Forest there is
minimum winter discharge and maximum discharge in autumn and spring. The Rhine
is thus a mixed rain and glacier river, with a more constant flow than the Maas (Table
8–5).

WATER SUPPLY IN THE RHINE-RUHR AREA AND
RANDSTAD HOLLAND

The demand for water in the Rhine-Ruhr area and the area of Randstad Holland
in the Rhine-Maas megapolis is partly active (i.e., consumable and industrial water
supplies and water for irrigation and outdoor recreation) and partly passive (i.e.,
water as an environmental factor in nature, agriculture, building — in lowland areas
shipping and fisheries).

Table 8–5 Discharge of the Rhine between 1901 and 1975, Projected for 2 × CO$_2$ in the Atmosphere, at Lobith (Dutch-German border) in m^3/sec

	1901-1975	2 × CO$_2$
Mean summer discharge	1850	1740
Minimum summer discharge	640	500
Maximum summer discharge	7150	6400
Mean winter discharge	2540	2920
Minimum winter discharge	620	750
Maximum winter discharge	13000	14000

From Deelstra, Tjeerd, Johan Galjaard, and Rob H. G. Jongman. *Possible Impacts of Climatic Change on European Cities.* The International Institute for the Urban Environment, Delft. 1989. With permission.

Table 8–6 Total Active Use of Water in North Rhine-Westphalia and the Netherlands

	North Rhine-Westphalia	Netherlands
Annual water-use (total)	6 × 10^9 m^3	
Annual water-use (no mining)	4 × 10^9 m^3	3.5 × 10^9 m^3
Groundwater	35%	29%
Surface water	65%	71%
Infiltration	48%	
Direct use	17%	

From Deelstra, Tjeerd, Johan Galjaard, and Rob H. G. Jongman. *Possible Impacts of Climatic Change on European Cities.* The International Institute for the Urban Environment, Delft. 1989. With permission.

Important problems in water availability that crop up regularly are the total available quantity, quality, and year-round distribution. Table 8–6 shows how water is used by the population of North Rhine-Westphalia and the Netherlands.

Mining activities alone account for one third of the water used in North-Rhine Wesphalia, where groundwater capacity is regionally restricted. Only in the northwest, the Nordrheinische Bucht, are there large aquifers. These are dwindling, however, due to lignite mining. The Rhine is an important source of drinking water for cities on the river.

Its tributary, the Ruhr, is the main source of drinking water in the area. Groundwater supplies are scant and too heavily polluted for the provision of drinking water. The Ruhr flow is subject to great fluctuation, and diminishes in dry periods. Reservoirs have been built in the Sauerland region to ensure sufficient water supplies in dry periods. Theoretically, dry periods of up to 6 months can be bridged. In practice the period is shorter, due to insufficient replenishment, sedimentation in the reservoirs, and the extra need for water.

Most drinking water in the Netherlands is extracted from groundwater. In the coastal provinces, where much of the Randstad is situated, natural groundwater is brackish. Water is therefore obtained from the Maas and Rhine. In 1984, 1334.2 million m^3 were extracted to meet the public requirements, 434.4 million m^3 of which were surface water. The respective expected amounts for 1995 are 1536.4 million m^3 and 465.4 million m^3.

In most regions there ought to be sufficient groundwater to meet requirements under normal circumstances. However, due to pollution and intensified draining in agriculture, problems occur. Particularly in dry periods, agriculture needs large amounts of water for irrigation and artificial raining and to prevent salination in western parts of the country. Public needs compete for the same amount of water.

In Belgium, Maas water is used not only by Liège and Charleroi, but also transported to Brussels and Antwerp. Groundwater supplies are already being used to the maximum, the requirements of large urban regions being increasingly supplemented by water for the Maas, transported by canals.

EXPECTED IMPACTS ON THE RHINE-RUHR AREA AND RANDSTAD HOLLAND

All over Europe, climatic change will cause higher temperatures, less summer precipitation and an increase in evapotranspiration. Winter precipitation may increase north of the Alps, but the period during which it is stored as snow will become shorter. Alpine glaciers will diminish; southward-flowing glaciers (toward Italy) will probably disappear.

North of the Alps, the hydrological cycle will be intensified by 5 to 10%, owing to:

- Increased precipitation, especially in autumn and winter
- Increased evapotranspiration in spring and summer

Mean winter temperature will rise by about 5% in southern Germany and the Netherlands, and mean summer temperature by 4°C and 3°C respectively. This means that effective precipitation in these parts will decrease, and with it available water supplies in summer. There is a mounting prospect of long, dry periods in spring and summer.

Climatic change will have two important effects on the Rhine-Ruhr area and Randstad Holland:

- Higher temperatures
- Reduced water supplies in summer

Tables 8–7 and 8–8 show possible climatic changes in the Rhine-Ruhr area and Randstad Holland. The simulated climate resembles that of Bordeaux and central France at present.

The river flow will react to the increase in winter precipitation and the decrease in effective summer precipitation with a growth of winter flow and a reduction of summer flow. Rivers with mainly rain-fed basins, like the Mosell, Maas, Main, and Ruhr, are highly sensitive to such changes. The upper Rhine is less susceptible, due to the greater input from snow-covered mountains. Even if the total annual discharge

Table 8–7 Expected Climatic Changes in the Rhine-Ruhr Area, Projected for $2 \times CO_2$ in the Atmosphere

	Observed	$2 \times CO_2$
Mean Jan. temp. (°C)	−1	5
Mean Jul. temp. (°C)	18	21
Decreased cloud in summer	—	−7%
Decreased soil moisture in summer	—	−40%

From Deelstra, Tjeerd, Johan Galjaard, and Rob H. G. Jongman. *Possible Impacts of Climatic Change on European Cities.* The International Institute for the Urban Environment, Delft. 1989. With permission.

Table 8–8 Expected Climatic Changes in the Randstad Holland, Projected for $2 \times CO_2$ in the Atmosphere

	Observed	$2 \times CO_2$
Mean Jan. temp. (°C)	+1	6
Mean Jul. temp. (°C)	17	21
Decreased cloud in summer	—	−9%
Decreased soil moisture in summer	—	−50%

From Deelstra, Tjeerd, Johan Galjaard, and Rob H. G. Jongman. *Possible Impacts of Climatic Change on European Cities.* The International Institute for the Urban Environment, Delft. 1989. With permission.

increases by 5%, the decrease in summer discharges will be significant. A tentative calculation for the Maas at Borgharen shows that the chance of a discharge of less than 1 m³ s in summer is increased by a factor of five to six.

In the case of the Rhine at Lobith, the far more variegated catchment and fast-melting glaciers will prevent extremely low summer discharges of the order calculated for the Maas. According to the same tentative calculation, however, the lowest summer flow will sink by 20% to 500 m³/s. The occurrence frequency of the lowest currently recorded summer discharge (640 m³) will increase by more than 10%. Winter discharge may also fluctuate more widely, due to the combination of increased precipitation and higher temperature, which causes a direct runoff of snow precipitation, or during more regular periods of thaw. The Rhine flow will be less regular than it is now. There will be an increase in high discharge in early spring, and of low discharge in August and September.

PROBLEMS IN COASTAL URBAN AREAS

Water supply problems can have a strong influence on living conditions in urban areas. North of the Alps, in central and western Europe, the problems will be less severe than in the Mediterranean region. However, the importance of water management and water planning will mount. Water shortages will be a normal problem, even in humid regions like the Rhine-Ruhr, Belgium, and the Netherlands, calling for a change in land use if water is to be conserved. Calculations of agricultural water requirements during a "10% dry year" and a temperature rise of 1.3°C show that the water in the Markermeer and Ysselmeer, the Netherlands' inland lakes, will only just meet the country's agricultural requirement. If other users need that water, and if the temperature goes up a few degrees, there will be grave problems.

In coastal regions the rising sea level, augmented by increased water discharge, will cause problems for tourist resorts and coastal urban zones. Erosion will increase along the Belgian and Dutch coast, causing problems, particularly for Belgian coastal urbanization. Increased river discharge in winter and early spring will force cities in northwestern Europe (Mannheim, Mainz, Coblenz, Cologne, Rotterdam) to make dikes and riverside walls higher. The cost of dike-raising on the lower branch of the Rhine, the river Waal, (0.6 to 0.9 m) this decade is estimated at 100 to 150 million Dutch guilders.

Winter tourism in the Alps will be hampered on the lower slopes. This means that tourism will move to higher regions, which are, however, more vulnerable. The Alpine tourist season will become shorter. Summer tourists are likely to move north, or to mountainous areas in central Europe or the C.I.S. The hot Mediterranean climate and scarcity of water will make summer holidays there a less attractive proposition. Higher temperatures will make countries on the North Sea and Ireland viable alternatives. The Mediterranean tourist season will be in winter.

CONSEQUENCES FOR THE RHINE-RUHR AREA AND RANDSTAD HOLLAND

Due to high precipitation and lake water shortage, Switzerland is unlikely to have a major water-supply problem. Possibilities for water storage are few and far between in southern Germany and the French Jura, where water supplies could pose a problem. In 1976, a dry year, 93% of the summer flow at Lobith was Alpine water.

Discharge from the French-German part of the basin was negligible in that dry year. Upper and middle Rhine discharge fluctuations have been increasing since the beginning of the century, due to river regulation. Water supplies could therefore become scarce in the upper Rhine section.

Improved climatic conditions favor living conditions in cities north of the urban regions of southern and central Europe. This means that current efforts in the Rhine-Ruhr area to compete with urban regions like Munich will benefit from the changing climate in the long term.

On the other hand, water supplies in the Rhine-Ruhr area and Randstad Holland will pose a great problem in summer. The Rhine-Ruhr area is at present using groundwater for industrial purposes, and mostly purified water from the Rhine for drinking. Mining activities and pollution are reducing the supply of groundwater. Rhine water supplies will dwindle much more in summer when the climate gets warmer. Its quality is already poor, and it is not yet clear how quickly it will improve.

People in the Ruhr area drink water from the Ruhr. Although under ideal storage conditions reservoirs can provide drinking water for a period of 6 months, the amount is less in practice. In the expected changed climate, evapotranspiration will increase in summer resulting in a 40% decrease in soil moisture. This means that a lot of water will vanish into the air, reservoir water too. If we had two or more dry years, water storage could prove inadequate, calling for measures to prevent shortages. Possibilities are

- The imposition of water-saving measures
- Water supplies from the North German Plain
- The construction of new reservoirs

In Randstad Holland, water from the Maas and Rhine has been used for the preparation of drinking water since 1985. Maas water is more polluted than Rhine water. In high-discharge periods Maas water is stored in artificial lakes. The province of South-Holland, where major parts of the Randstad are located, uses 290×10^6 m^3 of water from the Maas every year. This corresponds to an average daily use of 0.8×10^6 m^3. If extracted directly from the Maas, discharge would diminish by 9 m^3/s. Theoretically this would mean that in dry summers all the discharge would be used for drinking water, leaving nothing for agriculture or shipping. Under the expected climatic conditions, dry periods in summer can be more frequent and last longer. The change in discharge makes summer water shortages more likely. This situation will be aggravated if Maas water is extracted for storage in Belgium and France as well.

At present, Rhine water is also being used for the production of drinking water in the province of North-Holland, where Amsterdam and other northern parts of the Randstad are situated. When the climate changes, the quantity of Rhine water will be able to meet this demand, in dry periods too. However, Rhine and Maas water is used more intensively in the upstream area, and if pollution is not reduced, the quality of the water will deteriorate and be more difficult to clean. A solution could be more intensive purification in the upstream part of the basin, and the creation of water reserves in high-discharge periods.

The Rhine is an important inland shipping route. Climatic change reduces ships' draught. At 1983 price levels, the average increase in transport costs was estimated at 15 million guilders a year for the section of the Rhine between Rotterdam and Lobith. In point of fact, losses amounted to 70 million guilders in extremely dry years (this only applies to transport; industrial costs caused by, say, shortages of stocks, are ignored). Transport problems are already more acute on the Maas, and stand to become much more serious.

If water transport continues to be at risk, new solutions will have to be sought. Possibilities are

- To switch from shipping to road and rail transport
- To maintain the Rhine's depth by canalization etc.
- To move industrial plants dependent on bulk transport downstream

OUTLOOK

The Rhine-Maas megapolis is one of the world's most important industrial and urban areas, heavily dependent on rivers and their water. Climatic change can improve living conditions in the urban region, especially in the northern part of the Rhine-Ruhr area and Randstad Holland. It can, however, also cause serious problems in water supplies and in the rivers' transport function.

Conflict is likely to occur between parties interested in using the water. An important task for regional, environmental, and hydrological planning will be to solve the problems caused by the hydrological effects of climatic change.

There is a need for careful planning of the water supply and demand in various parts of Europe. Water-use planning will be more vital than ever on a supranational, national, and regional scale. This applies not only to the Rhine-Maas megapolis but also to other major transboundary rivers in Europe.

CONCLUSIONS

There are reasonable grounds to doubt society's willingness to radically alter consumption patterns and production processes, a necessary step if climatic change is to be avoided. Experts predicted the current environmental problems years ago, but failed to influence policy. Even now, when ways and means of coping with a host of environmental problems are being developed, policy moves sluggishly. It makes sense, then, to supplement studies on how to tackle the causes of climatic change with an examination of the possible consequences and methods of coping with them.

When more is known about possible consequences for cities, anticipatory measures at the level of urban planning can be taken in time. In some cases technical procedures will be involved, in others land use will be affected: for instance controlled migration to other areas more fit to live in.

It will also be easier to foresee that certain cities, for instance ports that no longer work properly, may lose their present economic functions. It will then be possible to anticipate the development of a different socioeconomic structure for the city.

However, combating the effects should be regarded as a second-best choice. Since the causes of the greenhouse effect are largely a product of the industrialized western world, it seems no less than logical to adapt technological processes in richer countries first. These countries are surely in a better position to bear the cost of adapting their cities as a consequence of climatic change than poor countries that are not yet or scarcely industrialized. Many of them are already stricken with drought, floods, ruined crops and associated disasters, and simply lack the means to cope with even more serious problems.

Continuing climate change will cause unprecedented migratory movements. And so concomitant problems will come on top of the ones resulting from local climate change. Cities capable of adapting to climatic changes by adopting urban planning measures will attract refugees and consequently face extra problems of a social and economical nature.

Urban planning will be confronted with the effects of rising temperature in various ways. First and foremost there are the existing urban structures, some of which will partly or wholly lose their function. Then there are the development schemes for the coming decades, many of which will prove to be inadequate in the long run. Finally, new cities will have to be planned, partly as a result of the changing climate. If the current trend continues, cities will have to change at a tremendous speed compared with the present planning rate. It may be possible to keep abreast of these developments from a technical point of view, but what about the effort and coordination needed for housing, physical planning, and building policy? It is quite conceivable that "squatter settlements" will form in the rich western world as a consequence of migration. Forget about urban planning in that case. The minute municipal authorities lose control of events, socioeconomic conditions will deteriorate and with them any chance of coping with climatic change. The resulting downward spiral will render the process all but irreversible. It is therefore vital to opt in the short run for a policy of greenhouse gas emission reduction, as well as to deal with the effects of climate change in cities.

REFERENCES

1. Nishioka, Shuzo, Yuichi Moriguchi, and Sombo Yamamura. Metropolis and climate change. The case of Tokyo. In: James McCulloch, Ed. *Cities and Global Change.* Climate Institute, Washington. 1991 (p. 115).

2. Deelstra, Tjeerd and Rob uit de Bosch. *The Greenhouse Effect. Preventative Urban Actions in the Netherlands.* The International Institute for the Urban Environment, Delft. 1993 (p. 19).

3. Gupta, Joyetta. A partnership between countries and cities on the issue of climate change, with special reference to the Netherlands. In: James McCulloch, Ed. *Cities and Global Change.* Climate Institute, Washington. 1991 (p. 69).

4. IPCC Working Group 2, Human Settlement; the energy, transport and industrial sectors; human health; air quality, and changes in ultraviolet-B radiation, *Climate Change: The IPCC Impact Assessment.* IPCC. 1990 (p. 5–9).

5. Deelstra, Tjeerd, Johan Galgaard, and Rob H. G. Jongman. *Possible Impacts of Climatic Change on European Cities. A Reconnaissance of a New Problem.* The International Institute for the Urban Environment, Delft. 1989 (p. 62).

6. Landsberg, Helmut E. *The Urban Climate?* Academic Press, New York, 1981; according to Annie Whiston Spirn. *The Granite Garden, Urban Nature and Human Design.* Basic Books Inc., New York. 1984 (p. 42).

7. Edgerton, Lynne. Warmer temperatures, unhealthier air and sicker children. In: James McCulloch, Ed. *Cities and Global Change.* Climate Institute, Washington. 1991 (p. 146).

8. Edgerton, Lynne. Warmer temperatures, unhealthier air and sicker children. In: James McCulloch, Ed. *Cities and Global Change.* Climate Institute, Washington. 1991 (p. 145).

9. Last, J. A., D. L. Warren, E. Pecquet-Goad, and H. P. Witsohl. Modification of lung tumor development in mice by ozone. In: *J. Natl. Cancer Inst.* 78, 149, 1987.

10. Waterman, R. E. Integrated coastal policy via building with nature. In: R. Frassetto, Ed. *Impacts of Sea Level Rise on Cities and Regions.* Proceedings of the First International Meeting 'Cities on Water' Venice, December 11–13, 1989. Marsilio Editori, Venice, 1991 (p. 64).

11. Turner, R. K., P. M. Kelly, and R. C. Kay. *Cities at Risk.* School of Environmental Sciences, University of East Anglia, U.K. (BNA International Inc., September 1990) (p. 35).

12. Turner, R. K., P. M. Kelly, and R. C. Kay. *Cities at Risk.* School of Environmental Sciences, University of East Anglia, U.K. (BNA International Inc., September 1990) (p. 33).

13. Hootsman, M. J. *Sea Level Rise.* A world-wide cost estimate of basic coastal defence measures. Delft Hydraulics Report H1068, (February 1990).

14. Gleick, P. H. and E. P. Maurer. *Assessing the Costs of Adapting to Sea-Level Rise. A Case Study of San Francisco Bay.* Pacific Institute for Studies in Development, Environment and Seanity, Berkeley, California, U.S.A. (April 1990).

15. Gleick, P. H. and E. P. Maurer. *Assessing the Costs of Adapting to Sea-Level Rise. A Case Study of San Francisco Bay.* Pacific Institute for Studies in Development, Environment and Seanity, Berkeley, California, U.S.A. (April 1990) (p. 38).

16. Gleick, P. H. and E. P. Maurer. *Assessing the Costs of Adapting to Sea-Level Rise. A Case Study of San Francisco Bay.* Pacific Institute for Studies in Development, Environment and Seanity, Berkeley, California, U.S.A. (April 1990) (p. 39).

17. Hootsman, M. J. *Sea Level Rise.* A world-wide cost estimate of basic coastal defence measures. Delft Hydraulics Report H1068, (February 1990).

18. Boorman, L. A., J. D. Goss-Custard, and S. McGrorty. *Climate Change, Rising Sea-Level and the British Coast.* Institute of Terrestrial Ecology, Natural Environment Research Council, London, Her Majesty's Stationary Office, 1989.

19. Enquete-Kommission 'Vorsorge zum Schutz der Erdatmosphäre des Deutschen Budestages (Hrsg.). *Schutz der Erde.* Bonn, Economica Verlag. Karsruhe, Verlag C. F. Müller. Teilband I. (p. 318).

20. Deelstra, Tjeerd, Johan Galgaard, and Rob H. G. Jongman. *Possible Impacts of Climatic Change on European Cities. A Reconnaissance of a New Problem.* The International Institute for the Urban Environment, Delft. 1989 (pp. 69–121).

21. Deelstra, Tjeerd, Johan Galgaard, and Rob H. G. Jongman. *Possible Impacts of Climatic Change on European Cities. A Reconnaissance of a New Problem.* The International Institute for the Urban Environment, Delft. 1989 (pp. 52–68).

CHAPTER 9

Impact of Climatic Change on the Ecology of Temperate Coastal Wetlands, Beaches, and Dunes

Vera Noest, Eddy van der Maarel, and Frank van der Meulen

INTRODUCTION

The possible impact of global climatic change, usually indicated as the greenhouse effect, on the coast and its ecology has caught the attention of the general public, mainly because of the fate of low-lying countries such as lowland Bangladesh or the Maldives facing a catastrophic rise in sea level. Even in countries such as the Netherlands with relatively strong protecting dunes there is much concern for the safety of the western part of the country. Apart from the immediate threat to lives and social organization there are many probable and possible changes in coastal ecosystems to be mentioned. Unfortunately, in most cases our assumptions about the general, i.e., global changes are still rather vague, and, moreover, the possible response of coastal ecosystems and landscapes can hardly be indicated because we lack historical data from which future changes could be derived. This chapter will treat some changes about which a prediction is not entirely unrealistic.

Among the effects put forward in the general discussions on global change the following processes are both of general and of direct interest within the framework of this contribution: an increase of the atmospheric CO_2 content, a general rise in air temperature, and a rise in sea level. From these general changes further changes can be derived, which will be discussed below. Estimations of the intensity or size of the global effects, and of their interrelations, vary largely between experts. In this contribution, global climatic changes are assumed to take place according to the predictions of the Intergovernmental Panel on Climate Change (IPCC).[1]

0-87371-301-X/95/$0.00+$.50

The main features are

- An increase in atmospheric CO_2 content to 450 ppm by the year 2050 and to about 520 ppm by the year 2100
- A rise in temperature of 1°C by 2025 and 3°C before the end of the next century
- A rise in sea level of 18 cm by 2030 and of 44 cm by 2070

Regional climate changes will differ significantly from these global mean values and predictions on a regional scale are much less reliable than the predictions on the global scale. Changes in regional precipitation (total amount and seasonal variation), wind velocity and direction, storm incidence, and the variability of the future climate (i.e., the incidence of extremes) are really hard to predict with a reasonable amount of confidence. However, these factors, in addition to the regional rise in temperature, will have a strong impact on the regional ecosystems, particularly on the interaction between abiotic and biotic processes.

For coastal ecosystems this list has to be extended with changes in storm incidence and changes in tidal range, in addition to a regional rise in sea level.

SPECIAL FEATURES AND MAIN TYPES OF COASTAL ECOSYSTEMS

Coastal ecosystems are all characterized by the proximity of the sea, the relatively strong influence of wind and airborne salt, and a maritime climate, i.e., a climate with smaller differences between summer and winter temperature than further inland. A second factor that distinguishes coasts from many other terrestrial ecosystems is their human use. In order to specify the special features of coastal ecosystems we will adopt a main division into types of coastal landscapes. In accordance with the structure developed in the series *Ecosystems of the World* we first distinguish between wet and dry coastal ecosystems,[2,3] which are further subdivided into:

- Salt marshes and mud flats
- Mangroves
- Coastal wetlands
- Coastal dunes and beaches
- Sea cliffs
- Raised reefs
- Skerry coasts

This contribution will concentrate on salt marshes and particularly on dunes and beaches. Effects on sea cliffs and skerries are presumably small, but the effects of a rising sea level on coastal wetlands may be large, especially through the loss of flat, low-lying land. The most extensive coastal wetland system is the coastal tundra, which borders much of the Arctic Ocean and the Canadian arctic waters. Indirect implications for the flora and fauna of this zone are difficult to appreciate. Regarding coastal wetlands in the tropics, we dare not say anything specific. Effects on raised reefs, and on coral reefs in general, will certainly be considerable, but this lies outside the expertise of the authors and, therefore, will not be treated here.

SALT MARSHES AND MUD FLATS

Salt marsh and mud flat systems widely occur in shallow seas or bays. They form mosaics and zonations of ecosystems, which are very important, both economically and ecologically. The main feature of a salt marsh is its zonation in relation to the tidal regime: a gradual transition from bare mud or sand flats at the line of low tide, through a variety of plant communities on the intertidal flat, to either a freshwater marsh or a dry terrestrial community. Each zone can be defined in terms of the frequency and duration of submergence, intensity of mechanical disturbance, and other components of the tidal regime. This often leads to a distinct zonation of communities parallel to the shore, although a more mosaic-like pattern can also be observed.

Salt marsh vegetation is characterized by low species diversity. It consists mainly of low-growing, salt-tolerant herbs, especially grasses. The prime colonizers of the lower flats are generally species of *Salicornia* and *Spartina*. Other typical genera with a global distribution include *Suaeda*, *Plantago*, *Juncus*, and *Puccinellia*.

The marsh is built up and grows seaward by the accretion of sediment. As the surface of the marsh rises, the frequency and duration of submergence decreases, which leads to a seaward shift of vegetation zones.

Impact of global change on distribution patterns will be mostly caused by temperature changes. There is a clear relationship between the isotherms of the coldest month and the geographical limits of salt marshes and mangroves: salt marshes occur in general in mid- and high latitudes, while mangroves are confined to the equatorial regions.[2] Maximum temperature rise (beyond the global mean) is predicted for high-latitudinal regions. This might well cause a shift in the distribution of mangroves from subtropical to temperate zones, notably along the eastern coasts of North America and Asia, and the southeastern coasts of South America, Africa, and Australia.

Regional salt marsh and mangrove systems will probably be affected most of all by changes in sea level and in tidal range, but such effects are still difficult to predict in detail. We will summarize here information on two well-known areas.

The Wadden Sea is one of the major salt marsh areas of the world, and well known through many studies (e.g., Reference 4). Moreover, some attention has already been given to the possible changes of the area as a result of global changes. We will therefore take this area as a case study.

Extensive salt marshes occur in the Wadden Sea area, both on the islands and on the mainland. Clear zonation patterns can be distinguished (see Reference 4 for historical notes) and on most of the Wadden islands transitions between the salt marshes and beach plains and dune slacks occur; they have a relatively high species diversity and harbor many rare plant species. The tidal flats contain few higher plants but huge numbers of lower plants (benthic diatoms) and animals, especially worms, molluscs, and crustaceans. Therefore, the area is extremely important as a nursery for many fish species. In addition, numerous bird species use the Wadden Sea as a breeding, feeding, moulting, and/or wintering area. Large parts of the world populations of species of ducks, geese, waders, gulls, and terns are largely dependent on the area.[5] Economically important are fisheries for shrimps and cockles, and aquaculture of mussels. Agricultural exploitation is mainly restricted to the mainland coast.

Present sedimentation rates on the island marshes in the Wadden Sea area are keeping up with the ongoing relative rise of sea level (0.44 cm/year over the last decades). A higher sea level rise of 0.5 to 1.0 cm/year may cause a reduction of the present island marsh area from 2800 to 450 ha.[6] Even if the marshes could compensate sea level rise by increased sedimentation, fluctuations in the tidal range are bound to induce changes in the vegetation composition. Olff et al. found distinct changes in species cover of the major species of the salt marsh of Schiermonnikoog, which could be correlated to fluctuations of the tidal inundation frequency.[7]

The Mississippi Delta has experienced a long-term relative sea level rise mainly due to regional subsidence, and offers valuable information about the impact of sea level rise on coastal wetlands. The current local rate of relative sea level rise is about 1.2 cm/year of which 85 to 90% is due to subsidence.[8,9] In the past several thousand years, the Mississippi Delta expanded considerably, despite the relative sea level rise.[10] During the last century, however, this accretion trend was reversed and land losses of 100 km²/year have been reported.[11] The main cause of these losses of coastal wetlands and marshes is not the relative sea level rise in itself, but the diminishing of sediment input.[12] Channeling, dam construction (for navigation and flood control purposes), and other management activities have effectively stopped both the influx of riverine sediment and the accumulation of resuspended sediment. Salt marshes that are cut off from their sediment source can no longer keep up with the relative sea level rise and are eventually drowned. Given sufficient input of sediment, both vertical and lateral expansion of the marshes may occur, even during periods of considerable sea level rise. Using a process-based spatial simulation model (CELSS), Constanza et al. computed a slight gain in total land area (10 km², or 0.7% of the total land area) for the Atchafalaya-Terrebonne marsh-estuarine complex, even when local rates of sea level rise were doubled to 0.46 cm/year (leading to a total rise of 1.03 cm/year, including subsidence).[13] The Mississippi River is at present changing from its current channel to the Atchafalaya River, thereby supplying the Atchafalaya-Terrebonne area with sufficient sediment. A second scenario, with a sea level rise of 1.67 cm/year (2.24 cm/year including subsidence) showed a net loss of land (19 km², or 1.3% of the total land area). These results are supported by earlier research by Baumann et al., who found that the marshes in the Atchafalaya area were able to maintain their surface elevation despite the ongoing subsidence and sea level rise.[14]

Other large coastal marsh systems we know of include the Camargue in southern France, the Marismas in the Coto Doñana National Park in southwestern Spain, the Banc d'Arguin off the Mauritanian coast, and the Mississippi Delta.[15-18]

Table 9–1 Human Influences Present in Coastal Areas Arranged According to the Ecosystem Component They Affect

Ecosystem	Human influence	Coastal defense	Water catchment-extraction	Water catchment-infiltration	Recreation-substrate	Recreation-landscape	Recreation-nature, research	Production of petroleum, gas	Extraction of sand	Collection of fruits	Forestry	Military training
Substrate	Appearance of artificial structures	X			X	X		X				X
	Accretion of sand	X										
	Removal of sand	X			X				X			
	Mobilization of sand				X				X			X
	Fixation of sand	X									X	
Soil structure	Ploughing up							X	X			X
	Compaction				X	X	X					
Soil water	Lowering phreatic water table		X									
	Raising phreatic water table			X								
	Inundation			X								
Nutrition	Eutrophication		X	X	X	X						
	Calcification				X				X			X
	Mineralization		X		X							
Plants	Removal	X			X	X				X		
	Introduction	X				X					X	
Vegetation	Removal	X			X				X			
	Treading				X	X	X					
	Planting	X									X	
Animals	Removal					X						
	Introduction										X	
	Disturbance				X	X		X				X

After van der Maarel.[20]

COASTAL DUNES AND BEACHES

GEOGRAPHICAL DISTRIBUTION AND SIGNIFICANCE

Dunes are even more important than salt marshes, both economically and ecologically. They occupy large areas along the shore all over the world and are particularly extensive along the Atlantic coasts of Europe, eastern and northwestern North America, southeastern South America and southeastern Africa, Japan, and most coasts of Australia.[2,3,19]

Probably the main function of dunes, at least for low-lying countries, is coastal defense. Several densely populated areas are protected from the sea by dunes (e.g., the Netherlands, southwestern France, eastern Spain, and the U.S. East Coast). Other important functions are water catchment and tourism. Table 9–1 lists the different types of land use of coastal dunes, together with the impact they have on different

Table 9–2 Main Gradients and Processes in Coastal Dune Ecosystems and the Major Climatic Factors Affecting Them

Component	Gradient	Process	Climate factor
Sand budget (foreshore)	Pos./neg.	Accretion/retreat	Sea level rise, marine currents
Sand budget (inland)	Pos./neg.	Erosion/accumulation	Wind climate, storm incidence
Moisture	Wet-dry	Humidification/desiccation	Precipitation groundwater level
Chloride	Fresh-salt	Desalination/salination	Salt spray, groundwater level
Carbonate content	Poor-rich in CaCO$_3$	Acidification/calcification	Sand budget, wind climate
Organic material	Humic-mineral	Decomposition	Temperature, moisture
Vegetation			Elevated CO$_2$, temperature, precipitation

After van der Maarel[21] and van der Meulen and van der Maare.[22]

ecosystem components.[20] Nature conservation as an alternative important land use is often in conflict with these land use functions.

GRADIENTS AND PROCESSES

Coastal dune ecosystems are characterized by the occurrence of many small-scale gradients, forming complexes. Table 9–2 lists the main gradients and processes acting on coastal dune ecosystems.[21,22] The fourth column lists the main aspects of climate change that will have a significant impact on these processes.

The gradients and processes in Table 9–2 may be considered in a hierarchical way[21,23] (Figure 9–1). Processes higher in the table are dominant over processes lower in the table. Vegetation will not only be influenced directly by climate change, but also by all processes on higher levels, either directly or indirectly.[24]

Influences of factors on lower levels on those on higher levels occur as well. An example on a small scale is the impact of rabbits on vegetation and on the soil surface (erosion). An example on the global scale is the fixation of CO$_2$ by plants and thus the influence of the phytosphere on the CO$_2$ concentration in the atmosphere.

In the original scheme of this hierarchy of spheres, the noosphere, the sphere of human activities, was placed lowest, because it is energetically dependent of the other levels. However, at the same time human activity has an ever-increasing ability to influence processes higher up in the hierarchy. The very problem of global change is the result of an essentially human impact on the atmosphere.

The following sections deal with the processes listed in Figure 9–1 in more detail.

COASTAL ACCRETION AND RETREAT

Sea level rise will in many places result in increased erosion and a trend toward landward movement of the coastline (transgression). This trend can be counteracted by reinforcement of natural (sand dune) or artificial (dike) barriers.[25] On sandy coasts,

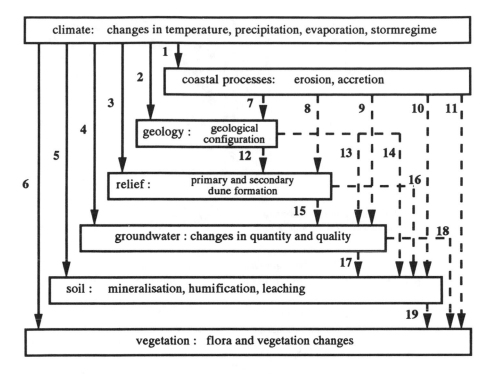

Figure 9–1 Main relations between landscape components, starting with climate. Feedback relations not shown. (After van der Meulen.[24])

however, the impact of sea level rise is largely dependent on the sand budget of the foreshore-beach area, with a negative sand budget leading to erosion of the beach and foredunes.[26] The formation of the recent dune coast of the Netherlands has taken place during a period of relative sea level rise, but with enough sediment being available.[27,28]

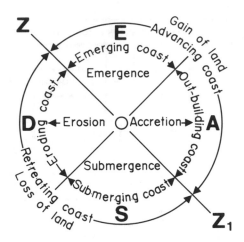

Figure 9–2 Valentin's scheme of coastal
classification. (After King.[29])

This dune landscape, the "Younger Dunes", was formed between 1000 and 1600, during a period of marine transgression, and partly covered the existing dune landscape. This former landscape, the "Older Dunes", was formed between 6000 and 3000 B.C. and can still be recognized on topographical maps by its elongated, parallel ridges with the troughs in between now filled with peat.

At emerging coasts, like in parts of the Gulf of Bothnia, sea level rise might be compensated by isostatic uplift. In other regions, however, such as the Wadden Sea area, the predicted sea level rise is superimposed on the relative rise of 15 to 20 cm per century as the result of isostatic lowering of the land.

Valentin's scheme of coastal classification (Figure 9–2) shows the combined effects of submergence or emergence and coastal accretion or erosion.[22,29] The line ZZ_1 represents the line along which two processes compensate each other.

Dune Erosion/Formation

A full vegetation cover of the dune surface will effectively control all erosion in the inner dune area. But where patches of bare sand occur at the surface, erosion processes can and will modify the geomorphology. Erosion by water tends to flatten the surface, while erosion by wind leads to the formation of both blowouts and accumulation dunes. Erosion by water can be far more important than by wind, measured in the amount of sand that is displaced.[30]

Moreover, Jungerius[31] found an intricate connection between the two processes: deflation patches are most likely to arise on the upper parts of slopes, after erosion by water has washed away the humic, water repellent sand and exposed the yellow sand underneath. Changes in the annual distribution of rainfall may influence the effects of water repellency of the sand. The effects of changing wind regimes on the development of blowouts is discussed by Jungerius et al.:[32] deflation of six blowouts in the Blink (the Netherlands) was found to be most highly correlated with SW winds with velocities between 6.25 and 12.5 m/s, which are the critical wind velocities for the particle sizes in the area (0.15 to 0.42 mm). Extreme events, e.g., NW storms, showed a tendency to fill the blowouts, which is interpreted as a first step of adaptation to a higher energy level. A shift in wind regime towards higher effective velocities may lead to a breakdown of the whole blowout-accumulation dune system, especially if this shift were accompanied by a change in wind direction.

Under the present climatological conditions, wind erosion very seldom leads to the formation of extensive deflation plains. Nearly all blowouts are stabilized spontaneously by either algae or vegetation;[33,34] hence, there is no need for expensive stabilization measures. If changes in wind, climate, and precipitation would enhance eolian erosion, species that can withstand considerable sand burial, such as *Ammophila arenaria* (marram grass) and *Salix arenaria* (creeping willow), would be favored. The survival of other pioneer plants in relation to increased sand burial will depend on the growth rate of such plants.

Changes in vegetation cover, caused by direct effects of climate change will probably be more important. When increased CO_2 levels lead to an increase in plant biomass and cover (see later discussion), this could reduce both wind and water erosion, and lead to a less dynamic dune landscape, both geomorphologically and ecologically. The interaction between abiotic and biotic processes in the dune environment then shifts toward a situation in which the biotic processes prevail.[24]

Humidification/Dessication

Underneath most dune areas lies a dome of fresh water, which is "floating" on top of the salt water. The fresh water table is raised above sea level due to the effective precipitation and the topography of the dunes.[23] A decreased width of the dune belt (as a result of a rising sea level) will in principle lead to a lowering of the water table in the remaining area. Sea level rise in itself has the opposite effect and will counteract the lowering of the water table.[35] Changes in annual precipitation have to be superimposed on this interaction, and even if total precipitation does not change, the effective precipitation may do so when vegetation cover and structure are changing.

For a coastal dune area with a length of 7 km in the Netherlands, Noest calculated net changes in groundwater level to vary from +36.5 to –21.0 cm, depending on the position with regard to the coastline and the amount of coastal retreat, assuming a sea level rise of 30 cm in 2050 and of 60 cm in 2090.[36]

Subsequently, a vegetation model was used to predict the expected species composition of 23 plots in the area. In the vegetation model a number of hydrological, climatological, and site variables were used to estimate the probability of occurrence of 100 dune valley species. Predictions were made for the 2 years under study (2050 and 2090) and for two management options: option 1 allows a coastline retreat of 25 m at most for 2050 and 50 m for 2090, and option 2 allows maximum coastline retreat of 50 m and 100 m, respectively. Canonical correspondence analysis was used to evaluate the changes in species composition of the plots.[37] With this ordination technique a number of independent (orthogonal) axes of maximal floristic variation are obtained, which are correlated with the environmental gradients included in the analysis.

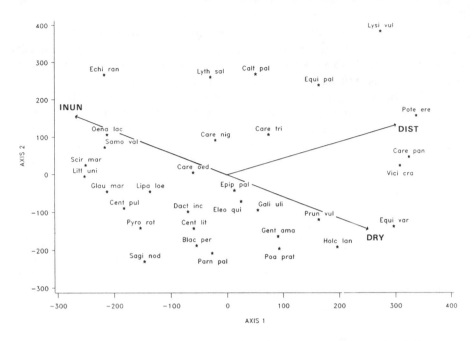

Figure 9–3 The position of the species centroids and the main environmental variables along the first two axes of maximal floristic variation from CCA. The projection of a species point onto the arrow of an environmental variable indicates the position of the species curve along the environmental gradient. See Appendix for abbreviations of species names. (From Noest, V. *Landscape Ecology* 6(1/2):89–97 (1991). With permission.)

Figure 9–3 shows 32 typical dune valley species and the 3 main environmental variables in an ordination diagram. The main variation in vegetation (on axes 1 and 2) is related to the variation in moisture as expressed by the parameters "duration of inundation" (INUN) and "duration of period with groundwater level >30 cm below surface" (DRY). A second gradient is based on distance to the foredunes (DIST).

Figure 9–4 shows the position of nine plots in the starting year (1989) and with the predicted vegetation in 2050 and 2090, for both management options. The connecting lines between the positions of one plot indicate the direction of change. The restricted coastline retreat under management option 1 leads for most plots to much wetter conditions, because the rise in phreatic level caused by sea level rise dominates the fall in phreatic level caused by the coastline retreat. There is a change in the direction of species occurring in wet valleys near the foredunes, which partly are indicators of brackish conditions. Under management option 2, the situation becomes reversed for many plots (notably plots 9, 13, 17, 12, and 10). Coastline retreat now dominates sea level rise, leading to drier conditions and a corresponding shift in species composition. The magnitude of this shift decreases from plot 9 toward plot 10 because of the increasingly inland positions of the plots.

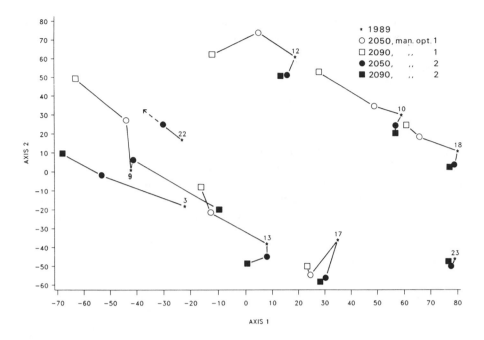

Figure 9–4 Sample scores (weighted average species scores) from CCA for nine plots. See main text for explanation. (From Noest, V. *Landscape Ecology* 6(1/2):89–97 (1991). With permission.)

Desalinization and Salinization

Airborne salt spray and intrusion of salty groundwater result in a higher chloride content of the dune sand. Salt spray is dependent on the speed and direction of the prevailing winds and the incidence of storms, and decreases exponentially with distance from the sea.[38,39] Coastal retreat will expose formerly inland parts of the dunes to increased salt spray, which may induce the return of salt-tolerant species. When the freshwater dome underneath a dune area is diminished by coastal retreat and sea level rise, the groundwater can become mixed with brackish water, especially in low-lying dune depressions (dune slacks, swales) behind the foredunes.

Acidification and Calcification

The current carbonate content of the superficial layers in a dune area is mainly a function of the initial lime content of the dune sand and the duration of leaching since the formation of the dunes. Decalcification of the top soil to a level of <0.3% $CaCO_3$ will lead to a sharp increase in pH, since the buffering capacity diminishes.[40,41] This acidification can be halted by the input of fresh sand with a high initial lime content. One source of fresh sand is the foreshore-beach system, but this only occurs when the sand budget is at least temporarily positive. A second source is the dune body itself: wind erosion and the formation of blowouts and secondary dune valleys will bring new, not yet decalcified sand to the surface.

Decomposition

The decomposition rate of surface litter and soil organic matter is largely controlled by temperature, soil moisture, and soil texture. In sandy, well-drained dune soils an increase in temperature and decrease in precipitation will lead to higher decomposition rates, lesser leaching, and lower acidification rates, at least in areas with a relatively large precipitation surplus.[42] In more arid regions, the hydrology of these soils may change from leaching to accumulative, while poorly drained soils may show a sharp increase in salinization. Where climate change leads to a significant change in composition of the vegetation and the litter produced, soil development may change drastically.

Vegetation Responses to Elevated CO_2 Concentrations

Climate change has both direct and indirect effects on plants. The direct effects include responses to increased atmospheric CO_2, higher temperatures, and changes in precipitation. Climate change will indirectly affect vegetation through changes in the above-mentioned processes, notably changes in moisture, nutrients, salt spray, and $CaCO_3$ content.

Elevated CO_2 levels have been shown to stimulate growth through enhanced photosynthesis, although this effect may decrease after some time, probably by starch accumulation and/or size constraints (self-shading).[43–45] Effects of elevated CO_2 concentrations will be physiologically different for the two main groups of photosynthetically active plants: C_3 and C_4 plants. The main difference between the groups lies in the biochemical pathways they use in photosynthesis. C_4 plants make very efficient use of the available CO_2, while CO_2 is a limiting factor for photosynthesis in C_3 plants. Effects of elevated CO_2 levels are therefore stronger on C_3 species than on C_4 species. Most coastal plants are C_3 species; examples of coastal C_4 species are *Salsola kali*, *Suaeda fructicosa*, and *Spartina townsendii*.

Increased CO_2 will reduce stomatal conductance, thereby increasing water use efficiency.[46] It may also lead to a decrease in stomatal density,[47] which may enable plants to withstand water stress slightly better than with present CO_2 levels.[48]

CO_2 enrichment of the atmosphere stimulates symbiotic nitrogen fixation in many agricultural legumes, as well as in some woody plants.[49,50] This can increase nutrient availability in infertile habitats. Especially in dunes, this can be an important factor.

However, most of the present knowledge stems from controlled experiments on single species.[51,52] Resource (light, nutrients) deficiencies are likely to modify the effect of CO_2 fertilization.[44,46] Moreover, the magnitude of the effects of elevated CO_2 concentrations differs considerably between species, thereby affecting species competition and, hence, community structure and species composition.[53]

Tracing increased biomass production may be difficult in many areas where the dunes are rich in nutrients, and/or are subjected to air pollution, notably the dry deposition of nitrogen compounds. In the relatively young dunes of Voorne (the Netherlands), near the big harbor and industry area of Rotterdam, which show a natural vegetational succession, a detailed vegetation map was made in 1959 and repeated in 1980.[54] Comparison of the maps and the accompanying floristic inventories showed an increase in nitrophilous and ruderal species and an overall trend towards a scrub/woodland complex. In other places — probably with more acid, leached soils — dry deposition may lead to dominance of grasses. Effects of succession, euthrophication by nitrogen deposition, and (future) CO_2 fertilization might be hard to quantify separately.

In summary, the main theoretical effect of increased CO_2 concentration is a larger total biomass, especially in dunes in temperate zones, where the C_3 pathway dominates. However, if there are limiting factors, notably moisture and nutrient availability, this effect may be much less in reality, while in nutrient- and humus-rich dune areas effects of a higher CO_2 level will be hard to separate from (natural) succession and eutrophication effects.

Vegetation Responses to Higher Temperature

The geographical distribution of vegetation types on a global scale is mainly defined by climatic variables. An obvious response to a temperature rise would be a general migration of plant species, including dominants, to higher latitudes and/or altitudes. However, past migration rates, especially of trees, have been much slower than the expected shift of isotherms (100 to 200 km poleward per degree of warming). Inadequate seed dispersal, a lag in soil development, and competition from persisting vegetation would force species to adapt to a different climatic regime.[55] Dispersal rates and adaptive abilities are extremely variable between species. Natural or man-made barriers to dispersal might cause local or even global extinction of some species (polar, montane, and island communities; isolated nature reserves), leading to a decrease in global species diversity.

The effects of elevated CO_2 on photosynthesis, stomatal conductance, and transpiration are generally enhanced by increasing temperatures.[43] A rise in mean temperature may cause insufficient hardening against colds and severe damage during incidental frosts. Insufficient winter chilling of buds may increase the required thermal time (accumulated day degrees >5°C) in some species (e.g., *Fagus sylvatica*), causing failure to take advantage of early springs. Species with low chilling requirements (e.g., *Crataegus monogyna*) may show early budburst and benefit from a prolonged growing season, albeit with an increased risk for subsequent frost damage.[56,57] Forest fire regime is closely correlated with climate and available fuel.[58] Higher temperatures, especially when combined with drought, will lead to an increase in fire frequency. When the forest understory is denser, due to direct CO_2 effects, fire intensities may also increase. This is likely to cause considerable changes in community structure and composition.

Vegetation Responses to Changes in Precipitation

An increased seasonality of precipitation, even when total annual precipitation remains constant, may result in prolonged flooding during winter and extensive periods of relative drought during summer, particularly in wet dune valleys. Most dune valley species are extremely sensitive to even small changes in groundwater level.[59]

Reduced summer precipitation, which may accompany rises in summer temperature, could well be the most important cause of change, the change being toward more open vegetation and a higher level of sand mobility. Temperate dune systems could become more semidesert-like. It is still impossible to say which species would be involved in such shifts of species composition.

Effects on Fauna

Few direct effects of changes in climate on mammals and birds of dune ecosystems are to be expected. They will, however, be affected by food availability and loss of habitat. Meekes has discussed the possible impact of climate change on bird species and concluded that of about 25 selected species only 3 or 4 may be threatened, whereas two species may even benefit from possible changes.[60]

Insect development, survival, growth, and reproduction increases linearly with temperature.[61] This may lead to increased frequencies of insect pests. Moreover, drought stress on plants tends to lead to more severe pest outbreaks (due to reduced chemical defense and increased nutrient levels).[62] Leaves of plants grown under elevated CO_2 conditions have a higher content of carbohydrates, resulting in lower leaf nitrogen concentrations and a higher C/N ratio.[63,64] This means that the nutritive value to herbivores is decreased, which can cause both increased feeding on C3 plants and reduced growth of insect herbivores.[65,66]

CONCLUSION

The overall conclusion is, not unexpectedly, that it is very difficult to predict any change, even a major one in coastal ecosystems as a result of global change. If sea level rise would occur to the extent predicted, this will doubtless cause the biggest overall changes, because both larger parts of salt marshes and low dune systems may disappear altogether. Changes as a result of increased temperature and CO_2 concentrations are much more difficult to predict and will depend very much on the species composition of the vegetation and the associated fauna composition.

REFERENCES

1. Houghton, J. T., G. J. Jenkins, and J. J. Ephraums, Eds. *Climate Change: The IPCC Scientific Assessment* (Cambridge, England: Cambridge University Press, 1990), p. 365.

2. Chapman, V. J., Ed. *Ecosystems of the World, Volume 1, Wet Coastal Ecosystems* (Amsterdam: Elsevier Scientific, 1977).

3. van der Maarel, E., Ed. *Ecosystems of the World, Volume 2A, Dry Coastal Ecosystems* (Amsterdam: Elsevier Scientific, 1993).

4. Dijkema, K. S. and W. J. Wolff. *Flora and Vegetation of the Wadden Sea Islands and Coastal Areas* (Leiden, The Netherlands: Stichting Veth, 1983), p. 413.

5. Smit, C. J. and W. J. Wolff. *Birds of the Wadden Sea* (Leiden, The Netherlands: Stichting Veth, 1980), p. 308.

6. Dijkema, K. S. Toekomstig beheer van kwelders op de eilanden en het vasteland, *Waddenbulletin* 26(3):118–122 (1991).

7. Olff, H., J. P. Bakker, and L. F. M. Fresco. The effect of fluctuations in tidal inundation frequency on a salt-marsh vegetation, *Vegetatio* 78:13–19 (1988).

8. DeLaune, R. D., W. H. Patrick, and R. J. Buresh. Sedimentation rates determined by Cs^{137} dating in a rapidly accreting salt marsh, *Nature* 275(5680):532–533 (1978).

9. DeLaune, R. D., R. H. Baumann, and J. G. Gosselink. Relationships among vertical accretion, coastal submergence and erosion in a Louisiana Gulf Coast marsh, *J. Sediment. Petrol.* 53(1):147–157 (1983).

10. Day, J. W., Jr. and P. H. Templet. Consequences of sea level rise: implications from the Mississippi Delta, *Coastal Manage.* 17:241–257 (1989).

11. Sasser, C. E., M. D. Dozier, J. G. Gosselink, and J. M. Hill. Spatial and temporal changes in Louisiana's Barataria basin marshes, *Environ. Manage.* 10(5):671–680 (1986).

12. Salinas, L. M., R. D. DeLaune, and W. H. Patrick, Jr. Changes occurring along a rapidly submerging coastal area: Louisiana, U.S.A., *J. Coastal Res.* 2(3):269–284 (1986).

13. Constanza, R., F. H. Sklar, and M. L. White. Modeling coastal landscape dynamics, *BioScience* 40(2):91–107 (1990).

14. Baumann, R. H., J. W. Day, Jr., and C. A. Miller. Mississippi deltaic wetland survival: sedimentation versus coastal submergence, *Science* 224:1093–1095 (1984).

15. Blondel, J. Les écosystèmes de Camargue, *Courr. Nat.* 35:43–56 (1975).

16. Valverde, J. A. An ecological sketch of the Coto Doñana, *Br. Birds* 51:1–23 (1958).

17. Moore, P., F. Garcia Novo, and A. Stevenson. Coto de Doñana. Survival of a wilderness, *New Sci.* 96:352–354 (1982).

18. Altenburg, W., M. Engelmoer, R. Mes, and T. Piersma. *Wintering Waders on the Banc d'Arguin Mauretania* (Groningen: Wadden Sea Working Group, 1982), p. 283.

19. Davis, J. L. *Geographical Variation in Coastal Development* (London: Longman, 1977).

20. van der Maarel, E. Environmental management of the coastal dunes in the Netherlands, in *Ecological Processes in Coastal Environments*, R. J. Jefferies and A. J. Davy, Eds. (Oxford: Blackwell, 1979), pp. 543–570.

21. van der Maarel, E. On the establishment of plant community boundaries, *Ber. Dtsh. Bot. Ges.* 89:415–443 (1976).

22. van der Meulen, F. and E. van der Maarel. Coastal defence alternatives and nature development perspectives, in *Perspectives in Coastal Dune Management,* F. van der Meulen, P. D. Jungerius, and J. Visser, Eds. (The Hague, The Netherlands: SPB Academic Publishing, 1989), pp. 183–195.

23. Bakker, T. W. M. *Nederlandse Kustduinen; Geohydrologie,* (Wageningen, Netherlands: Pudoc, 1981), p. 189.

24. van der Meulen, F. European dunes: consequences of climate change and sea level rise, in *Dunes of the European Coasts,* T. W. Bakker, P. D. Jungerius, and J. A. Klijn, Eds. (Cremlingen-Destedt, Germany: CATENA Verlag, 1990), pp. 209–223.

25. Gerritsen, F. Impact of climatic change on coastal defense, in *Climatic Change: Impact on Coastal Habitation,* D. Eisma, Ed. (Chelsea, MI: Lewis Publishers, Inc., in press).

26. van der Meulen, F., J. V. Witter, W. Ritchie, and S. M. Arens. Precepts, approaches and strategies, *Land. Ecol.* 6(1/2): 7–13 (1991).

27. Jelgersma, S., J. d. Jong, W. H. Zagwijn, and J. F. v. Regteren Altena. The coastal dunes of the western Netherlands: geology, vegetational history and archaeology, *Medede. Rijks Geol. Dienst N.S.* 21:93–167 (1970).

28. Zagwijn, W. H. The formation of the Younger Dunes on the west coast of the Netherlands (AD 1000–1600), *Geol. Mijnbouw* 63:259–268 (1984).

29. King, C. A. M. *Beaches and Coasts,* (London: Arnold, 1972), p. 570.

30. Jungerius, P. D. and F. van der Meulen. Erosion processes in a dune landscape along the Dutch coast, *Catena* 15:217–228 (1988).

31. Jungerius, P. D. Geomorphology, soils and dune management, in *Perspectives in Coastal Dune Management,* F. van der Meulen, P. D. Jungerius, and J. Visser, Eds. (The Hague, Netherlands: SPB Academic Publishing, 1989), pp. 91–98.

32. Jungerius, P. D., J. V. Witter, and J. H. v. Boxel. The effect of changing wind regimes on the development of blowouts in the coastal dunes of The Netherlands, *Land. Ecol.* 6(1/2):41–48 (1991).

33. Ancker, J. A. M., P. D. Jungerius, and L. R. Mur. The role of algae in the stabilization of coastal dune blowouts, *Earth Surf. Proc. Land.* 10:189–192 (1985).

34. Pluis, J. L. A. and B. de Winder. Natural stabilization, in *Dunes of the European Coasts,* T. W. Bakker, P. D. Jungerius, and J. A. Klijn, Eds. (Cremlingen-Destedt, Germany: CATENA Verlag, 1990), pp. 195–208.

35. Carter, R. W. G. Near-future sea level impacts on coastal dune landscapes, *Land. Ecol.* 6(1/2):29–39 (1991).

36. Noest, V. Simulated impact of sea level rise on phreatic level and vegetation of dune slacks in the Voorne dune area (the Netherlands), *Land. Ecol.* 6(1/2):89–97 (1991).

37. ter Braak, C. J. F. Canonical correspondence analysis: a new eigenvector technique for multivariate direct gradient analysis, *Ecology* 67:1167–1179 (1986).

38. Sloet van Oldenruitenborgh, C. J. M. and E. Heeres. On the contribution of air-borne salt to the gradient character of the Voorne Dune area, *Acta Bot. Neerl.* 18:315–324 (1969).

39. Vulto, J. C. and P. J. M. van der Aart. Salt spray and its influence on the vegetation of the coastal dunes of Voorne and Goeree (The Netherlands) in relation to man-made changes in coastal morphology, *Verh. K. Ned. Akad. Wet.* 2–81:65–73 (1983).

40. Rozema, J., P. Laan, R. Broekman, W. H. O. Ernst, and C. A. J. Appelo. On the lime transition and decalcification in the coastal dunes of the province of North Holland and the island of Schiermonnikoog, *Acta Bot. Neerl.* 34(4):393–411 (1985).

41. Grootjans, A. P., H. Esselink, R. v. Diggelen, P. Hartog, T. D. Jager, B. v. Hees, and J. O. Munninck. Decline of rare calciphilous dune slack species in relation to decalcification and changes in local hydrological systems, Coastal Dune Congress 1989, Sevilla, Spain.

42. Sevink, J. Soil development in the coastal dunes and its relation to climate, *Land. Ecol.* 6(1/2):49–56 (1991).

43. Cure, J. D. and B. Acock. Crop responses to carbon dioxide doubling: a literature survey, *Agric. For. Meteorol.* 38:127–145 (1986).

44. Lenssen, G. M., J. Lamers, M. Stroetenga, and J. Rozema. Interactive effects of atmospheric CO_2 enrichment, salinity and flooding on growth of C_3 (*Elymus athericus*) and C_4 (*Spartina anglica*) salt marsh species, *Vegetatio* 104/105:379–388 (1993).

45. Poorter, H. Interspecific variation in the growth response of plants to an elevated ambient CO_2 concentration, *Vegetatio* 104/105:77–97 (1993).

46. Warrick, R. A., H. H. Shugart, M. J. Antonovsky, J. R. Tarrant, and C. J. Tucker. The effects of increased CO_2 and climatic change on terrestrial ecosystems, in *The Greenhouse Effect, Climatic Change and Ecosystems*, B. Bolin, B. R. Döös, J. Jäger, and R. A. Warrick, Eds. (Chichester: John Wiley & Sons, 1986).

47. Woodward, F. I. Stomatal numbers are sensitive to increases in CO_2 from pre-industrial levels, *Nature* 327:617–618 (1987).

48. Morison, J. I. L. and R. M. Gifford. Plant growth and water use with limited water supply in high CO_2 concentrations. II. Plant dry weight, partitioning and water use efficiency, *Aust. J. Plant Physiol.* 11:375–384 (1984).

49. Hardy, R. W. F. and U. D. Havelka. Photosynthate as a major factor limiting nitrogen fixation by field-grown legumes with emphasis on soybean, in *Symbiotic Nitrogen Fixation in Plants*, P. S. Nutman, Ed. (Cambridge, England: Cambridge University Press, 1976).

50. Norby, R. J. Nodulation and nitrogenase activity in nitrogen-fixing woody plants stimulated by CO_2 enrichment of the atmosphere, *Physiol. Plant.* 71:77–82 (1987).

51. Woodward, F. I. Global change: translating plant ecophysiological responses to ecosystems, *Trends Ecol. Evol.* 5:308–311 (1990).

52. Arp, W. J., B. G. Drake, W. T. Pockman, P. S. Curtis, and D. F. Whigham. Interactions between C_3 and C_4 salt marsh plant species during four years of exposure to elevated atmospheric CO_2, *Vegetatio* 104/105:133–143 (1993).

53. Reekie, E. G. and F. A. Bazzaz. Competition and patterns of resource use among seedlings of five tropical trees grown at ambient and elevated CO_2, *Oecologia* 79:212–222 (1989).

54. van der Maarel, E., R. Boot, D. van Dorp, and J. Rijntjes. Vegetation succession on the dunes near Oostvoorne, the Netherlands; a comparison of the vegetation in 1959 and 1980, *Vegetatio* 58(3):137–187 (1985).

55. Davis, M. B. Lags in vegetation response to greenhouse warming, *Climatic Change* 15:75–82 (1989).

56. Cannell, M. G. R. and R. I. Smith. Climatic warming, spring budburst and frost damage on trees, *J. Appl. Ecol.* 23:177–191 (1986).

57. Murray, M. B., M. G. R. Cannell, and R. I. Smith. Date of budburst of fifteen tree species in Britain following climatic warming, *J. Appl. Ecol.* 26:693–700 (1989).

58. Clark, J. S. Effect of climate change on fire regimes in northwestern Minnesota, *Nature* 334:233–235 (1988).

59. van der Laan, D. Spatial and temporal variation in the vegetation of dune slacks in relation to the ground water regime, *Vegetatio* 39:43–51 (1979).

60. Meekes, H. T. H. M. The possible impact of climate change on the avian community of dune ecosystems, *Land. Ecol.* 6(1/2):99–103 (1991).

61. Ratte, H. T. Temperature and insect development, in *Environmental Physiology and Biochemistry of Insects*, K. H. Hoffman, Ed. (Berlin, Germany: Springer Verlag, 1984).

62. Rhoades, D. F. Herbivore population dynamics and plant chemistry, in *Variable Plants and Herbivores in Natural and Managed Systems,* R. F. Denno and M. S. McClure, Eds. (New York: Academic Press, 1983).
63. Overdieck, D. and F. Reining. Effect of atmospheric CO_2 enrichment on perennial ryegrass (*Lolium perenne* L.) and white clover (*Trifolium repens* L.) competing in managed model-ecosystems. I. Phytomass production, *Acta Oecologica/Oecologia Plantarum* 7:357–366 (1986).
64. Overdieck, D. and F. Reining. Effect of atmospheric CO_2 enrichment on perennial ryegrass (*Lolium perenne* L.) and white clover (*Trifolium repens* L.) competing in managed model-ecosystems. II. Nutrient uptake, *Acta Oecol./Oecol. Plant.* 7:367–378 (1986).
65. Lincoln, D. E., D. Couvet, and N. Sionit. Response of an insect herbivore to host plants grown in carbon dioxide enriched atmospheres, *Oecologia* 69:556–560 (1986).
66. Fajer, E. D. The effects of enriched CO_2 atmospheres on plant-insect herbivore interactions: growth responses of larvae of the specialist butterfly, *Junonia coenia* (*Lepidoptera: Nymphalidae*), *Oecologia* 81:514–520 (1989).

APPENDIX: LIST OF SPECIES ABBREVIATIONS AND NAMES

Species of young, wet slacks near the foredunes (partly brackish):

Cent pul	Centaurium pulchellum
Echi ran	Echinodorus ranunculoides
Glau mar	Glaux maritima
Litt uni	Littorella uniflora
Oena lac	Oenanthe lachenalii
Samo val	Samolus valerandii
Scir mar	Scirpus maritimus

Species of open, calcareous slacks:

Blac per	Blackstonia perfoliata
Care oed	Carex oederii
Cent lit	Centaurium littorale
Dact inc	Dactylorhiza incarnata
Eleo qui	Eleocharis quinqueflora
Epip pal	Epipactis palustris
Gali uli	Galium uliginosum
Gent ama	Gentianella amarella
Lipa loe	Liparis loeselii
Parn pal	Parnassia palustris
Pyro rot	Pyrola rotundifolia
Sagi nod	Sagina nodosa

Species of older slacks with a well-developed moss and humus layer:

Calt pal	Caltha palustris
Care nig	Carex nigra
Care tri	Carex trinervis
Equi pal	Equisetum palustre
Lysi vul	Lysimachia vulgaris
Lyth sal	Lythrum salicaria

Species of relatively dry slacks, with a mown vegetation:

Care pan	Carex panicea
Equi var	Equisetum variegatum
Holc lan	Holcus lanatus
Poa prat	Poa pratensis
Pote ere	Potentilla erecta
Prun vul	Prunella vulgaris
Vici cra	Vicia cracca

CHAPTER 10

Impact of Climatic Change on Coral Reefs, Mangroves, and Tropical Seagrass Ecosystems

Alasdair J. Edwards

INTRODUCTION

In the previous chapter the potential impacts of climatic change on the ecology of temperate coasts were addressed. Beaches, dune systems, and rocky coasts are likely to respond in broadly similar ways to climatic change whether they lie in cold temperate regions or in tropical seas. However, certain ecosystems are restricted to tropical coastlines or are particularly well developed there and it is these ecosystems — coral reefs, mangroves, and seagrasses — on which this chapter concentrates.

Before discussing the potential impacts of different aspects of climatic change on each of these ecosystems in turn, the predicted magnitude and time scale of the climatic changes that will be used as the basis for discussion need to be defined. Due to the major uncertainties surrounding predictions, discussion of impacts will focus on responses to best-guess estimates of climatic change.

The simplistic approach of separately examining the responses of each ecosystem (and even then of only the dominant organisms) to each of several different manifestations of climatic change (e.g., sea level rise, temperature rise, CO_2 increases, etc.) in turn is adopted to gain a first-order view of what are potentially the most significant impacts for each ecosystem. Two points must be borne in mind at all times: first, the different aspects of climatic change *do not* act separately but in concert, and second, the three ecosystems being discussed *do not* exist in isolation but are often highly interdependent. The first point means that negative impacts and the resultant stresses on organisms in the systems may be cumulative (or even worse, synergistic), while beneficial impacts (such as the CO_2 fertilization effect on plants)

0-87371-301-X/95/$0.00+$.50
© 1995 by CRC Press, Inc.

may antagonize certain negative impacts (e.g., thermal stress). The second point means that a factor causing a negative impact on one ecosystem may, by its damage to that system, indirectly result in negative impacts on a second ecosystem. Unfortunately, the considerable uncertainty about the responses of ecosystems to single aspects of climatic change and our limited understanding of the interactions between the systems prevent any meaningful consideration of the holistic response of tropical coastal ecosystems. However, potentially significant interactions between systems and interactions between their responses to different aspects of climatic change will be highlighted where appropriate.

SCENARIOS FOR GLOBAL CHANGE WITH PARTICULAR ATTENTION TO IMPACTS ON TROPICAL COASTAL ECOSYSTEMS

Scenarios for global climatic change are in continual flux as underlying assumptions used to estimate greenhouse-gas emissions (e.g., predictions of population growth, economic growth, energy supplies, progress on political agreement on emission controls, rates of deforestation, etc.) are refined, continuing scientific research leads to better understandings of key processes, and models (e.g., global circulation models — GCMs) are improved. Predictions forming the basis for discussion in this chapter center on the Intergovernmental Panel on Climate Change (IPCC) 1992 Scenario A (IS92a), one of six alternative scenarios offered by IPCC in 1992,[1] which replace the IPCC 1990 Scenario A (known generally as the Business-As-Usual scenario or SA90[2]) with a range of potential "Business-As-Usual" futures. IS92a lies in the middle of the range of 1992 scenarios for greenhouse-gas emissions (expressed as CO_2 equivalents) and is the 1992 scenario most similar to SA90.[3] Using this scenario and models that have incorporated the negative radiative forcing effects of sulfate aerosols and negative feedback resulting from enhanced plant biomass productivity caused by increasing CO_2 concentrations (CO_2 fertilization effect), best-guess projections of radiative forcing and consequent global-mean temperature rise and sea level rise have all been revised downwards recently by Wigley and Raper.[4] Their predictions using IS92a suggest a best-guess global-mean warming of 2.5°C (range 1.7 to 3.8°C) and a best-guess global-mean sea level rise of 48 cm (range 15 to 90 cm) by the year 2100. This corresponds to a 0.2°C warming per decade and a rise in sea level of 4 mm/year. Over the past 100 years global-mean temperature has risen by about 0.45 ± 0.15°C and sea level by about 10 to 15 cm.[5,6] Thus, rates of warming are predicted to be about five times those that have occurred over the last century and rates of sea level rise about four times faster. It is these accelerated rates of change, unprecedented since the dawn of civilization, which give the greatest cause for concern for mankind. Now over a billion people live on tropical coasts and in one way or another have lifestyles dependent on coastal ecosystems. One positive effect of increased carbon dioxide (the CO_2 fertilization effect) is that global net primary production may increase by 20 to 26% for a doubling of CO_2.[7]

Predictions of global change are subject to marked uncertainty as demonstrated by Monte Carlo simulations of sea level rise (Figure 10–1) and at present useful predictions

Monte Carlo Probability Distributions of Sea Level Rise

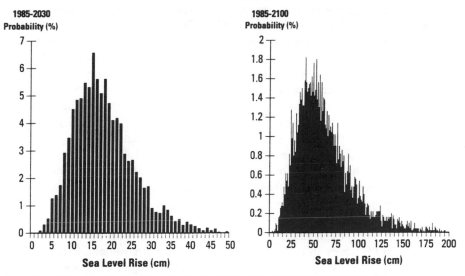

Source: Titus and Narayanan, "The Probability Distribution of Future Sea Level Rise," in The Rising Challenge of the Sea,
proceedings of the IPCC's Coastal Zone Management Subgroup, Margarita Island, Venezuela, in press.

Figure 10–1   For the purposes of this chapter best-guess predictions are used; however, the
levels of uncertainty should not be forgotten. For example, these Monte Carlo
probability distributions for sea level rise (taken from Reference 94) based on
SA90 indicate that for a best-guess rise of about 15 cm by 2030, there is a 5%
chance of a rise of less than 6 cm or greater than 36 cm, and a 50% chance of
a rise of less than 10 cm or greater than 20 cm. For planners, in particular, such
levels of uncertainty create major problems.

of climatic change on a regional scale are on the whole not possible.[8] However, studies
of past climate fluctuations and GCM predictions both suggest that warming in tropical
regions will be less than the global-mean warming.[5,9] There is much debate about how
much less this may be. Holdgate et al.[10] suggest that the tropics will warm by 0.7 to 0.9
times the global average, which would indicate a best-guess warming for the tropics of
perhaps 2°C by 2100. However, paleoclimatic studies of the warm periods of the
Holocene Optimum (5000 to 6000 years B.P.) and the Eemian interglacial (125,000 to
130,000 years B.P.) indicate that despite average summer temperatures in high northern
latitudes 3 to 4 and 4 to 8°C, respectively, warmer than today, there was no detectable
warming in the tropics.[5] Use of palaeoclimates as analogs of future warming must be
treated with caution because of the different forcing factors involved — changes in
seasonal distribution of solar radiation vs. globally averaged rise in greenhouse gas
concentrations. However, there is independent (but controversial) support for limitations
on warming in the tropics centering around possible feedback mechanisms (involving
formation of highly reflective cirrus clouds), which come into play when sea-surface
temperature exceeds 30°C and keep sea-surface temperatures below 33 to 35°C.[11,12] For
the purposes of this chapter a best-guess estimate of a 2°C rise in tropical mean sea-
surface temperature by 2100 will be adopted.

Models also appear to be in reasonable agreement that areas which experience heavy monsoon-type rains may experience even more intense local rainstorms at the expense of gentler but more persistent rainfall.[9,13] This has important implications for marine ecosystems such as fringing coral reefs and seagrass beds sensitive to bursts of freshwater, nutrient, and sediment runoff. Tropical storms (hurricanes, typhoons, and tropical cyclones) form only where sea-surface temperatures are 26 to 27°C or greater, which might lead one to expect their more widespread occurrence in a warmer climate.[9] However, many other factors influence storm formation and whether or not tropical storms may increase (or decrease) in frequency or intensity, or both, is still controversial[5,9,14] despite a widespread public view that they will increase in frequency and intensity, and indeed already have. The IPCC 1990 view that, although rising sea-surface temperatures may increase the theoretical maximum intensity of tropical storms with a possible 40 to 50% increase in their destructive potential for a doubling of atmospheric CO_2,[15] "... climate models give no consistent indication whether tropical storms will increase or decrease in frequency or intensity as climate changes" still seems to prevail. However, some recent work suggests there may be an increase in frequency in tropical storms on doubling atmospheric CO_2 (cited in Reference 14). The potential impact of possible increased episodic storminess will not be considered further except to say that coral reefs, mangroves, and seagrass ecosystems all help to protect tropical coastlines from storms, and that the economic damage caused by storms is likely to increase if these ecosystems are adversely impacted by climatic change or local anthropogenic disturbance.

IMPACT OF CLIMATIC CHANGE ON CORAL REEFS

Coral reefs tend to fringe coasts where winter monthly mean temperatures are above 20°C. They are thus present throughout most of the tropics. On the western margins of the oceans where the thermocline tends to be deep and where warm currents extend tropical conditions to almost 60° of latitude, coral reefs tend to be best developed. By contrast, on the eastern margins where the thermocline may be only 40 m deep, upwelling common, and where cold currents flowing equatorward and squeeze tropical waters into about 30° of latitude, coral reefs tend to be poorly developed (eastern Pacific) or more or less absent (eastern Atlantic). The principal center of diversity of corals is in the Indo-Pacific around Indonesia and the Philippines with a subsidiary and relatively species-poor Atlantic center of diversity in the Caribbean.

Coral reef ecosystems are dominated by scleractinian and milleporid corals, which build most of the reef framework along with less obvious but nonetheless significant calcareous coralline red algae. Carbonate sediments produced by the breakdown of these and other organisms, notably the calcareous green alga *Halimeda*, provide infill for the reef framework and contribute to coral reef accretion over time. Coral reef ecosystems are thus dominated by calcifying organisms that are depositing about 0.6 to 0.9 Gt of $CaCO_3$ globally each year.[16] The key to rapid rates of calcification in reef-building scleractinian and milleporid corals lies in the symbiotic association between the coral polyps and the vegetative stages of unicellular brown dinoflagellate algae called zooxanthellae, which can live in the cells lining the coral polyps' internal digestive cavities. During daylight, photosynthesis by the zooxanthellae produces carbon compounds for the host coral animal and helps to mop up CO_2 produced by the calcification process. The net effect is to increase the rate of calcification by two to three times. Because the rate of calcification and, thus, reef accretion is light-dependent, corals grow best in clear shallow waters with maximum rates of reef accretion between 5 and 15 m depth.[17] Zooxanthellate corals can survive at depths greater than 60 m but grow very slowly in deep water. Greatest coral species diversity tends to be found at 10 to 30 m depth.

In looking at the impacts of climatic change on coral reefs we therefore need to consider responses to potential change of not only the coral animals themselves, but also of their symbiotic algal partners. The intimacy of the symbiosis, unfortunately, makes it difficult to disentangle these responses.

The impacts of global change — including non-climatic anthropogenic disturbances of global scale — have recently been the subject of an excellent review by Smith and Buddemeier,[18] which is commended to the reader wanting a more detailed account. Impacts of global change on coral reef ecosystems and priorities for future research have also been discussed in depth at an international workshop in Miami in 1991.[19] The global climatic impacts of significance to coral reefs are likely to be increases in sea temperature, sea level rise, and possibly increasing levels of ultraviolet radiation, although the latter is not strictly climatic. Locally or regionally, changes in storm patterns and currents may have impacts on coral communities while changes in rainfall patterns may have local effects on sedimentation, incidence of salinity anomalies, and nutrient inputs to fringing reefs. In this section only potential responses of coral reefs to the global impacts will be considered.

SEA LEVEL RISE

The first question to be investigated is whether rates of vertical accretion of coral reefs can keep up with the predicted rate of eustatic sea level rise, the best guess for which is 4 mm/year. Buddemeier and Smith[20] reviewed available information on reef accretion and concluded that the best overall estimate of the sustained maximum rate of vertical accretion was 10 mm/year. Thus, if healthy, reef ecosystems as a whole are clearly in no danger of being drowned by sea level rise; indeed, since the last ice age they have survived rates of rise of 20 mm/year during two separate periods lasting roughly 1000 years each.[21,22] However, both natural and anthropogenic disturbance can profoundly influence their ability to respond to sea level rise. For example, a Panamanian reef, which, before suffering over 50% coral mortality as a result of sea surface warming during the 1982–1983 El Niño-Southern Oscillation event, was depositing about 10 tonnes $CaCO_3 ha^{-1}y^{-1}$ is now eroding at approximately 2.5 tonnes $CaCO_3 ha^{-1}y^{-1}$ which is equivalent to vertical erosion of 6 mm/year.[23]

From mankind's viewpoint one of the most important services provided by coral reefs is coastal protection, and it is not theoretical maximal accretion rates that are the issue. Of more interest are the results of a method for evaluating reef growth known as the alkalinity depression technique,[24] which sheds light on the likely modal rates of accretion of reefs in different environments. This technique indicates that shallow lagoonal reef systems (where water circulation may be restricted) have modal upward accretion rates of only 0.6 mm/year, while coral reef flats have rates around 3 mm/year and areas of coral thickets rates of about 7 mm/year.[20] Thus, even the relatively slow accreting reef flats, which absorb wave energy offshore and thus serve to protect coastlines, should on average be able to grow vertically almost as fast as the best-guess rate of sea level rise, other factors being equal (Figure 10–2). These reef flats may benefit from sea level rise as their upward growth is currently constrained by exposure to air at extreme low tides leading to limitation of growth of corals and survival of only hardier species. Frequency of such detrimental exposure may be expected to decrease with accelerated sea level rise and, thus, reef-flat accretion rates may be enhanced and perhaps the diversity and community structure of reef-flat communities also affected positively.[16,18] Coral communities within semi-enclosed lagoonal systems, where temperature and salinity tend to fluctuate more widely than in the open ocean and where nutrient depletion can occur, may also benefit from improved water circulation as a result of sea level rise.[18,24]

TEMPERATURE

Corals in tropical seas live at temperatures that during the summer months are close to their upper lethal limits.[25] These lethal limits appear to vary geographically and to be correlated with normal summer ambient maximum sea temperatures. Thus, in the Arabian Gulf corals on offshore reefs survive at 34°C for weeks or months[26] and some species on inshore reefs can survive prolonged exposure to temperatures of 36 to 38°C, while corals from Hawaii would be expected to suffer mass mortality if exposed to temperatures around 31°C for prolonged periods.[27] Corals at Enewetak in the central Pacific where normal summer maxima are about 2°C higher than at Hawaii have an upper lethal limit that is also about 2°C higher.[25] Thus, over long time scales (millennia) reef-building corals and their symbiotic zooxanthellae have adapted to a range (albeit fairly narrow) of ambient temperature regimes and in some parts of the world are already adapted to temperatures well above those to which most corals are likely to be exposed over the next century. However, whether at a given location they can adapt to an unprecedented 2°C rise in mean sea surface temperature over a century is highly debatable.

Living so close to their upper lethal limits, corals are very sensitive to small anomalous rises in sea temperature such as occur during El Niño events. These can cause breakdown of the photosynthetic pigments of the symbiotic zooxanthellae and loss of symbiotic zooxanthellae themselves (coral "bleaching") and, if the anomaly is large or prolonged, subsequent death of coral colonies.[28–30] Exposure for only a few days to temperature elevations of 3 to 4°C above normal summer ambient maxima, or for several weeks to elevations of only 1 to 2°C above summer ambient can cause coral bleaching.[25,27,31] Depending on both the time of exposure and magnitude of the temperature elevation, the corals may either recover as remaining zooxanthellae

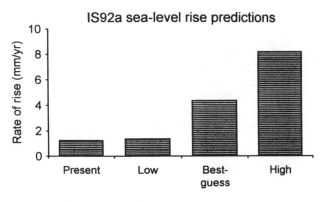

A

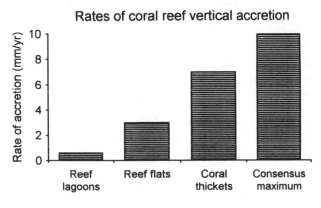

B

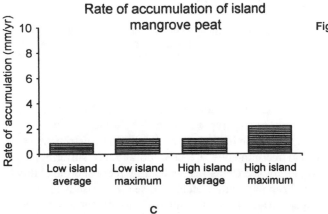

C

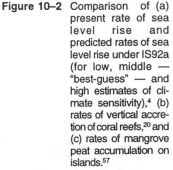

Figure 10–2 Comparison of (a) present rate of sea level rise and predicted rates of sea level rise under IS92a (for low, middle — "best-guess" — and high estimates of climate sensitivity),[4] (b) rates of vertical accretion of coral reefs,[20] and (c) rates of mangrove peat accumulation on islands.[57]

rebuild chloroplasts and multiply asexually, or die. Under field conditions, massive bleaching followed by 90 to 95% mortality has been recorded for temperature rises of 4 to 5°C for periods of 1 to 2 days[27] while low mortality followed by rapid recovery

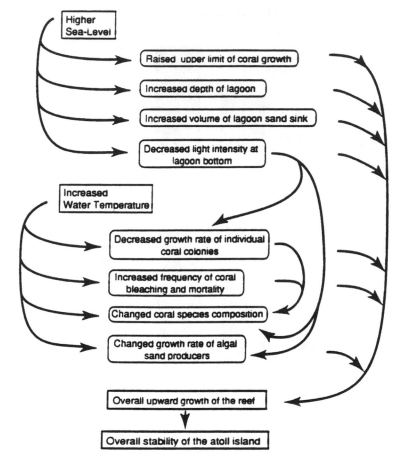

Figure 10–3 Interaction of various factors in determining vertical accretion of atoll reefs and reef-based islands. (From J. C. Pernetta and D. L. Elder, 1992, Climate, sea level rise and the coastal zone: management and planning for global changes, *Ocean & Coastal Management.* With permission.)

has been reported following 1 to 2°C temperature elevations. Bleaching markedly reduces coral growth and severely impairs reproduction.[32,33] Further, temperature changes even smaller than those that cause bleaching are known to be detrimental to coral growth, reproduction, and recruitment.[27] Clearly, coral reefs suffering the effects of temperature rise would not accrete as fast as healthy reefs and the ability of temperature-stressed reefs to keep pace with accelerated sea level rise is likely to be seriously impaired.[23,24] Figure 10–3 shows the interactions of various factors determining upward growth of atoll reefs.

Thus, on one hand, we already have coral bleaching and associated stress responses[35] being triggered by unexpected departures from normal summer ambient maximum temperatures taking place over a time scale of days to weeks, while on the other we have the prospect of a rise in global-mean sea temperature, of a similar magnitude to these departures, taking place over centuries. Two separate but related issues thus require consideration: (1) will corals and their zooxanthellae be able to adapt to a slow rise in mean sea-surface temperatures of slightly less than 0.2°C per decade? and (2) will warming anomalies become more frequent or more intense, causing increased stress of coral reefs and leading to (a) reduced growth, reproductive success, and recruitment of corals, (b) more frequent and more severe bleaching events, and (c) higher mortality rates of corals?

As the temperature of the oceans increases so it seems likely will the incidence of events that would *today* be regarded as anomalous. However, at a particular locality a rise in sea temperature that would today be considered as a 2°C rise above normal summer ambient maximum temperatures (sufficient to cause bleaching), is likely in the year 2100 to be regarded as a normal summer ambient maximum temperature. There does not appear to be evidence that there will be increase in climatic variance consequent on warming,[2] so the magnitude of temperature anomalies should remain fairly constant about a rising mean. Thus, the two questions resolve into the single question of whether corals can adapt fast enough for today's anomaly to become tomorrow's norm as far as coral/zooxanthellae physiologies are concerned. At present there is little evidence either way and this is clearly an area requiring further research since it is fundamental to predictions of coral reef community response to climatic change. What little evidence there is suggests that local adaptive change does not take place over periods of several years.[36]

The severity of the likely detrimental effects of warming on coral reefs will depend on the degree of lag between adaptive responses and sea temperature rise, but disruption of coral reef ecosystems seems inevitable as we move from an era (over 5000 years) of relative climatic stability to one of change. The magnitude of this lag will be related to the generation times of corals and their zooxanthellae with the expectation that the zooxanthellae will be able to adapt to change far more quickly than their hosts.

Different species of corals, colonies of the same species in different locations, and different clones of single species at one location have been shown to have differing thermal sensitivities.[25,27,36,37] The raw material for selection of thermal tolerance thus appears to be available, although there are doubts (based on research on insects) about the ability of existing species to adapt successfully to high temperature stress particularly if associated with other stresses.[38] In any event, as changing selective pressures come to bear, coral reef community structure could be significantly altered and reef accretion seriously impaired with detrimental effects on coastal erosion and on those people dependent on coral reefs for their livelihoods. For example, branching corals such as species of *Acropora* and *Pocillopora* are among the most thermally sensitive;[18] such species harbor bait-fish used for tuna pole-and-line fishing and their demise could thus have significant effects on local fishing communities.

ULTRAVIOLET RADIATION

Predicted changes in surface ultraviolet radiation levels are not really climatic changes but are global in impact and will be briefly outlined here. Stratospheric ozone absorbs shorter wavelength (<315 nm) solar ultraviolet radiation and prevents the most damaging wavelengths (<286 nm) from reaching the earth's surface at all. Near-ultraviolet radiation (around 290 to 400 nm) does reach the earth's surface and its shorter wavelengths (up to 320 nm — often referred to as UVB) can cause skin cancer in humans and damage to planktonic organisms such as fish and crustacean larvae.[39] Depletion of the ozone layer caused largely by chlorofluorocarbon compounds (notably CFC-11 and CFC-12 used in aerosols, refrigeration equipment, and foam manufacture) increases surface exposures to UVB with effects being most marked at high latitudes.[40] Depending on how effective curbs on CFCs and other ozone-depleting chemicals are, following the Montreal Protocol initiatives, it is predicted that globally averaged total column ozone may decrease by about 1 to 4% by 2035.[41] Since percentage increase in surface UVB radiation is a little over double the percentage decrease in total column ozone,[39] average surface UVB exposures in the tropics may be expected to increase by about 2 to 8% above today's levels.

Little is known about what effects such increases in surface UVB radiation will have. Ultraviolet radiation penetrates clear seawater well,[42] but is absorbed within about 5 cm of the surface in turbid waters, e.g., on some fringing reefs in Southeast Asia. Some corals produce "sun-screen" compounds (e.g., mycosporine-like amino acids isolated from S-320 pigment[43]). These give broadband protection from the most biologically damaging incident wavelengths of ultraviolet radiation, absorbing UV at wavelengths of 285 to 350 nm with peak absorbance at around 320 nm.[43,44] The sensitivity of corals to UV radiation appears to be directly related to the depth at which they occur.[45] Further, the amount of UV-absorbing substances produced is inversely correlated with depth down to 20 m and thus appears to be directly correlated with levels of incident UV radiation.[43] Also, certain strains of zooxanthellae are more UV-tolerant than others.[44] Thus, it seems that mechanisms for adaptation to the small changes in surface UVB radiation predicted are in place and that the impacts of increased UVB radiation are likely to be secondary to other impacts of climatic change. However, UVB can be expected to exacerbate stresses on shallow water coral reef communities as production of UV-absorbing compounds appears to be energetically expensive[46] and the molecules may be relatively short-lived. Also, concentrations of UV-absorbing substances have been shown to decrease at higher temperatures[47] indicating that elevated temperature could antagonize coral responses to increased UVB.

CORAL REEFS AND GLOBAL CARBON BUDGETS

Coral reefs are estimated to cover about 600,000 km^2 of the earth's surface[48] (0.17% of the ocean surface). Assuming regional averages for gross calcium carbonate production of 1 to 1.5 kg $CaCO_3$ m^{-2}y^{-1}, an estimate of global calcium carbonate production of 0.6 to 0.9 Gt $CaCO_3$ y^{-1} can be derived.[49,50] During the Holocene something in the order of 9000 Gt of coral reef limestone may have been laid down.[16] Intuitively one would thus think that coral reefs, by precipitating $CaCO_3$ and sequestering carbon, would act as a sink for atmospheric CO_2. This is not the case on the time scales of decades to centuries in which we are examining potential impacts of climatic change.[18] On the contrary, the calcification process actually generates CO_2 as shown in the simplified equation below:

$$Ca^{2+} + 2HCO_3^- \leftrightarrow CaCO_3 + H_2O + CO_2$$

As $CaCO_3$ is precipitated by the calcifying organisms of the reef, the seawater becomes more acidic due to the removal of bicarbonate and carbonate ions (despite its buffering powers which results in only 0.6 mol of CO_2 being produced per mole of $CaCO_3$ deposited), reducing the solubility of CO_2 and increasing its partial pressure (pCO_2) in the seawater.[50] The pCO_2 of surface waters of the oceans is only very slightly less than that in the atmosphere (<10 µatm) and increases in pCO_2 in the water column will lead to escape of CO_2 to the overlying atmosphere.

In the short term (hours to days) much of the CO_2 generated by calcification may be absorbed by photosynthesis but in the longer term (seasons or longer) coral reefs may contribute about 0.02 to 0.08 Gt C as CO_2 annually to the atmosphere[50] (a similar amount to that released from the oil well fires in Kuwait in 1991[1]). This is not very significant when considered as a percentage (0.3 to 1.3%) of anthropogenic emissions of CO_2 from burning of fossil fuels, which amounted to a contribution of C in 1989 and 1990 of about 6.0 Gt/year. Since the oceans are overall a sink for CO_2,[51] coral reefs are perhaps best considered as reducing global uptake of C by oceanic surface waters by about 0.02 to 0.08 G/year.

As atmospheric CO_2 doubles from about 360 ppm now to over 700 ppm by 2100 (based on IS92a[3]), so will the pCO_2 of oceanic surface waters. This will markedly decrease the saturation states of aragonite (the form of $CaCO_3$ generally deposited by reef-building corals and calcareous green algae such as. Halimeda) and other carbonates and possibly have deleterious effects on calcification.[18] The increase in atmospheric CO_2 also has the potential to increase the growth of C_3 plants such as some macroalgae and seagrasses by stimulation of photosynthesis, depressing respiration, and relieving nutrient and high-temperature stresses (the CO_2 fertilization effect).[52] At the coral reef ecosystem level this possible reduction in calcification on the one hand, and enhancement of photosynthesis on the other, could cause a shift from coral-dominated to turf- and fleshy algae-dominated communities on the reef.[18] Any major shift away from an ecosystem dominated by calcifying organisms to one dominated by non-calcifying algal species and bioeroders (as has occurred on some eastern Pacific reefs badly affected by the 1982–1983 El Niño[23,24]) would clearly severely compromise the ability of reefs to keep pace with sea level rise as well as having profound effects on reef fish and other communities of importance to humans. Until more research is done the likelihood of such scenarios must remain anyone's guess.

IMPACT OF CLIMATIC CHANGE ON MANGROVES

Mangrove ecosystems or "mangals"[53] are upper intertidal tropical ecosystems dominated by a taxonomically diverse (34 species from nine families) group of salt-tolerant trees called mangroves.[54] Mangrove species tend to have very specific require-ments for growth and this results in zonation of species within the intertidal zone. Mangals have considerable value as spawning and nursery grounds for commercially exploited fish and prawns, natural breakwaters protecting coastlines from erosion, a filter for terrestrial runoff trapping silt and slowing release of nutrients and freshwater into coastal waters, a source of a wide range of products from wood to honey, refuges for rare fauna, and for educational and tourism uses.[55] As for corals, there are two centers of diversity for mangroves, the Indo-Malayan region and the Caribbean.[54]

Mangrove species mostly occur between mean sea level and mean high water spring tide on sheltered shores of the tropics. The poleward limits of their range appear to coincide with minimum mean monthly sea-surface temperatures of around 15°C but close to these limits only one or two mangrove species are found and these are generally stunted in growth.[56]

Mangals tend to be most diverse and extensive on continental margins or on the shores of large high islands where they often benefit from allochthonous inputs of sediments (usually terrigenous in origin — Figure 10–4). Examples are the Sundarbans area of the Ganges and Brahmaputra Deltas in Bangladesh, the Florida Everglades, the deltas and estuaries of northern and northeastern Australia, the Niger Delta, and the coasts of Papua New Guinea, Borneo, and Sumatra. Some such mangals, by trapping sediments and mangrove tree litter in their root systems, have prograded seaward at rates of tens of centimeters or more per year. On low-lying coral island shores (e.g., Maldives, Grand Cayman, and Tongatapu) and on arid coastlines (e.g., Red Sea and Arabian Gulf) where terrigenous inputs of sediments are lacking, rates of accumulation of litter and sediment are much slower.[57] Dating and other studies of cores of mangrove sediments have allowed the rates of accumulation during the Holocene and positions of past mean tidal levels to be estimated.[58–60] These rates indicate how fast mangals are able to accrete vertically and, thus, their likely response to accelerated sea level rise.

SEA LEVEL RISE

During the early Holocene when sea level was rising fast (up to 20 mm/year) as the polar ice sheets receded after the last glaciation, large mangrove swamps did not exist. It was only when sea level stabilized around 6000 years B.P. that extensive mangrove ecosystems became reestablished and then only in sheltered locations as the coral reef flats that protect many coastlines today would not yet have devel-oped.[21,57] While sea level was rising rapidly mangroves probably survived as narrow coastal fringes and scattered stressed trees migrating landward with each generation as the intertidal zone shifted.

Studies of mangrove peat cores from low island areas without significant inputs of terrigenous sediments indicate average vertical accretion rates of only 0.8 mm/year with some indication that such mangrove systems have in the past kept up with rates

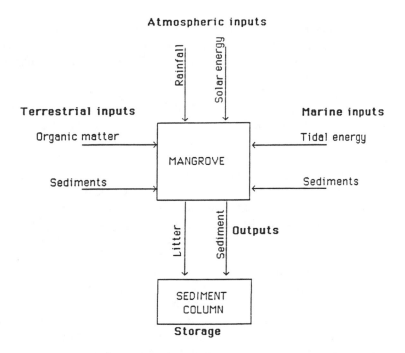

Figure 10–4 Fluxes within the mangrove ecosystem that contribute to the sediment column and thus the ability of mangrove to keep pace with accelerated sea level rise. (From J. C. Ellison, Possible effects of sea level rise on mangrove ecosystems, paper presented at the Small States Conference on Sea Level Rise, Malé, 14–18 November, 1989. With permission.)

of sea level rise of 1.2 mm/year.[57] These mangals could thus be coming under stress at current rates of eustatic sea level rise (depending on local tectonic movements).* On high islands (e.g., western Samoa, Fiji, Kosrae) where there can be significant terrigenous input of sediments, accretion rates of up to about 2.2 mm/year have been recorded.[61,62] Thus neither low nor high island mangals (Figure 10–2) appear to be able to accrete vertically as fast as the best-guess rate of sea level rise of 4 mm/year. Also as sea level rises, so will the shoreline erosion tend to increase[63] with increased removal of detritus and sediment from the mangal edge by waves. Mangrove stands in such situations will thus tend to be compressed and may suffer reductions in species diversity.

Certain species of mangrove tree are likely to be more robust in the face of accelerated sea level rise than others. In the Caribbean, Vicente et al.[64] note that the aerial roots of the black mangrove (*Avicennia germinans*) project only short distances above intertidal muds while prop roots of the red mangrove (*Rhizophora*

* Recent work in the Florida Keys suggests this scenario is pessimistic. The *increase* in mangrove area of 3.6 to 8.0% per decade has been recorded on low lying Key islands concomitant with relative sea level rise of 2.3 mm/year. [Snedaker, S. C., Meeder, J. F., Ross, M. S., and Ford, R. G. Discussion of Ellison, Joanna, C. and Stoddart, David, R., 1991. Mangrove ecosystem collapse during predicted sea level rise. Halocene analogues and implications. *Journal of Coastal Research*, 7(1), 151–165, *J. Coastal Res.* 10:497–498 (1994)].

mangle) stand much higher above mean sea level (sometimes more than 6 m). They therefore suggested that in the relative short-term rising sea level may cause massive local extinctions of black mangrove but have insignificant effects on red mangroves. Also, Ellison,[57] working in Tonga, considered *Rhizophora* communities likely to be more robust to rising sea level over a time scale of many generations having found peat accumulation rates (during a period of falling sea level when rates are higher) beneath *Rhizophora* (5.3 mm/year) to be double those beneath the black-mangrove species *Bruguiera* and *Excoecaria* (2.6 mm/year).[57]

In areas where extensive mangals exist in deltaic plain settings with large inputs of terrigenous sediment (often as a result of unwise agriculture or forestry in upland areas), such as the Sundarbans area of Bangladesh, they are likely to be more resilient to accelerated sea level rise (putting to one side, for the moment, such factors as local subsidence). However, it should be noted that too much allochthonous sediment being washed to the coast as a result of poor agricultural practice can bury pheumatophores (breathing roots) of mangroves, killing them.[57] Pernetta and Osborne[65] suggested that the extensive Purari Delta mangrove systems in the eastern Gulf of Papua, which is actively accreting at present, may retreat landwards but suffer little reduction in extent even for a 2-m rise in sea level. They contrast this with a section of the Gulf of Papua to the west where tidally dominated estuaries are dependent on longshore drift for supplies of sediments to maintain coastal stability. Here they suggest that sea level rise may reduce sediment supply (marine input of Figure 10–4) and lead to a recession of the coastline, compression of the mangrove zonation, and reduction in its extent by over 60%.

Pethick and Spencer,[66] working in the 1022 km² Rufiji Delta in Tanzania, point out that seaward erosion of mangroves may provide a source of sediments that can promote accelerated vertical accretion of the mangrove swamp landward (analogous to the situation in temperate salt marshes). Their preliminary results indicate that the seaward fringe of the Rufiji Delta may be eroding horizontally at up to 40 m/year but vertical accretion may at the same time be occurring within the mangrove system at up to 20 mm/year. Thus, although some compression is occurring and there is severe loss of the seaward strip of *Avicennia*, the prognosis for the bulk of the mangrove system may not be as bad as might at first appear. The rate of tectonic subsidence of the deltaic sediments and thus the local rate of sea level rise is, unfortunately not known for the Rufiji Delta. This leads to the widespread problem of deltaic subsidence, which means that many of those areas with the best developed mangals and with allochthonous supplies of sediment to sustain them may already be subsiding faster than predicted rates of eustatic sea level rise.

The Bengal Delta formed by the confluence of the Ganges, Brahmaputra, and Meghna Rivers and fringed by the huge Sundarban mangroves transports over 1000 million tons of sediment annually to the Bay of Bengal, but does not appear to have prograded significantly in 200 years.[67] A subsidence rate of 10 mm/year would account for about half the terrigenous inputs, the rest presumably accumulating on the inner shelf or escaping to the deep sea.[67] This estimate of the rate of subsidence is probably conservative, but nonetheless, would require the Sundarbans mangroves to keep pace with a local rate of sea level rise of at least 14 mm/year. Thus, where mangroves are potentially most robust to sea level rise they may have to cope with the highest local rates of rise because of tectonic effects.

TEMPERATURE, CO_2, AND ULTRAVIOLET RADIATION

Rises in sea-surface and air temperatures in tropical and subtropical regions by about 2°C by 2100 are likely to be beneficial to mangrove species living near the poleward limits of present-day distributions leading to increased species diversity, greater litter production and larger trees in these peripheral mangrove systems. There is also potential for limited poleward range extensions of mangroves. The possibility of thermal stress affecting mangrove root structures and establishment of mangrove seedlings at sea temperatures above 35°C has also been noted.[68]

Mangroves are C_3 plants and thus should also benefit significantly from increased atmospheric CO_2 concentrations with increased productivity. Despite living in wet conditions mangroves have remarkably low transpiration rates and thus, high water-use efficiencies. This adaptation may be linked to problems of salinization of the soil around the roots as water is taken up because of the exclusion of most salt at the root surface.[69] Increased CO_2 may allow even lower transpiration rates and reduce potential salinization problems. Overall, these ameliorating circumstances are likely to be of little importance compared to the impact of accelerated sea level rise.

Leaves of some mangrove species contain phenolic compounds, which have UV-absorbing properties and have been shown to be protective against the damaging effects of UVB radiation. There is also some evidence that UV radiation may stimulate the production of these compounds in some *Bruguiera* and *Rhizophora* species.[70] Thus, they seem to have the potential to adapt to the small increase in incident UVB, which is predicted.

IMPACT OF CLIMATIC CHANGE ON TROPICAL SEAGRASSES

Although concerns about the effects of long-term climatic change and more immediate phenomena such as coral bleaching have stimulated considerable interest and research into potential impacts on coral reef ecosystems and to a lesser extent mangals, relatively little appears to have been published on likely impacts on other tropical coastal ecosystems. Thus, potential impacts on seagrass ecosystems can be discussed only briefly and speculatively.

Seagrass ecosystems are dominated by monocotyledonous grass-like flowering plants from two families that have successfully adapted to inter- and subtidal marine environments. There are about 50 species worldwide.[71] Seagrasses are found from about midtidal level to up to 70 m deep in clear water[72] but tend to be best developed in shallow waters (<10 m deep).[73] Seagrasses, like mangroves, can trap sediments and can thus accrete vertically and in some areas form substantial mounds raised tens of centimeters to meters above the surrounding sediments.[74] Rates of vertical accretion do not appear to have been much studied but a figure of 10 mm/year is quoted for Mediterranean *Posidonia* beds.[75] The broad leaves of larger species (e.g.,

Thalassodendron, Enhalus, Thalassia) act as baffles, slowing water movement, promoting sediment deposition, and reducing its resuspension by wave action.[74,76] They thus can have significant indirect and direct coastal protection functions, stabilizing unconsolidated sediments with their extensive rhizome and root systems, reducing wave energy reaching the shore with their leaf canopies, and also reducing sediment loads reaching coral reefs toward the sea.

Seagrasses are also of economic importance in other ways.[77] They provide shelter for postlarvae and juveniles, and habitats and food for herbivorous commercial fish species (parrotfishes, surgeonfishes, and rabbitfishes) and for the prey of carnivorous fishes caught commercially.[73,77] Further, they provide the principal habitat for some commercial prawn fisheries, for example, fisheries for *Penaeus semisulcatus* in the Arabian Gulf and *P. duorarum* off Florida, and shelter for postlarvae and juveniles of other commercially fished prawns, and of spiny lobsters and crabs of economic value.[73,78] Seagrasses are also a principal food of Green turtles (*Chelonia mydas*), dugongs, and manatees,[79] and in some regions provide important feeding grounds for some species of waders and other shore birds.[73] Dense seagrass meadows are among the most productive of marine ecosystems with net productivities of C reaching 100 $gm^{-2} y^{-1}$.[80] Establishment of seagrass communities effectively turn vast areas of unconsolidated sediments, ranging from fine muds to coarse sands, with relatively low productivity and uniform relief into highly productive plant-dominated structured habitats (Figure 10–5). Off the Gulf of Mexico coast of Florida alone it is estimated that there are 8500 km^2 of seagrass beds.[73]

It seems likely that intertidal and shallow water seagrasses (<5 m depth) are most likely to be impacted by climatic change. Plants living at depth and limited by the available light will also no doubt suffer eventually from increasing water depth, but in such situations net productivity will be very low and plants small and scattered. For seagrasses there seem to be few generalizations about impacts of climatic change that can be made, with likely community responses being very dependent on local factors, not the least of which may be responses of adjacent coral reefs and mangroves.

TEMPERATURE

Like corals, seagrasses, particularly tropical species, appear to be relatively stenothermal and although able to survive short exposures to temperatures higher (or lower) than the ambient, are adversely affected by prolonged departures from the ambient.[73] In Florida optimum productivity of the turtle grass *Thalassia* occurs at 28 to 30°C and short-term exposure (i.e., over a tidal cycle) to 33 to 35°C seawater appears to be tolerated. However, sustained exposures to 33 to 35°C, or shorter exposures to 35 to 40°C leads to leaf loss, inhibition of root growth, and eventually death.[64,73,80] In general it appears that sustained elevations of sea temperatures of 4 to 5°C above summer ambient maxima will cause extensive mortality. A mean sea-surface temperature rise of 2°C by 2100 will clearly exacerbate the stress of seagrasses already living near their thermal tolerance limits in shallow lagoons with poor circulation or near thermal effluents of power plants (e.g., in Puerto Rico and Florida).[64] The magnitude of the impact of mean sea-surface temperature rise will

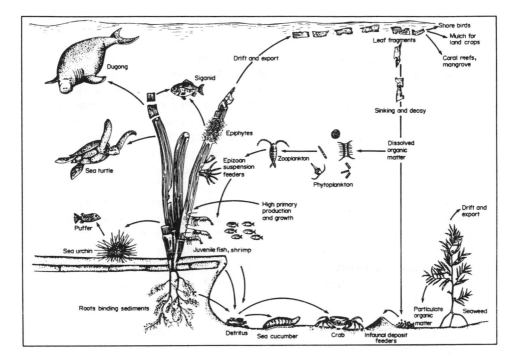

Figure 10–5 Food chain in Philippine seagrass ecosystems showing the potential uses of the plants and their functions in the coastal environment. (From M. D. Fortes, 1989, *Seagrasses: A Resource Unknown in the ASEAN Region*. With permission.)

depend on whether seagrasses can adapt fast enough to the rising temperature. The temperate eelgrass *Zostera marina* seems to exist as a series of ecological races adapted to different local temperature regimes and with varying upper thermal limits.[81] Potentially, as global-mean temperature rises the geographical distribution of these races may shift. The disadvantage to tropical seagrasses currently at the limits of thermal tolerance is that no races pre-adapted to the higher temperatures predicted (unprecedented over millions of years), exist.

It has been suggested that small increases in sea temperature over a number of years affected North Atlantic *Zostera* communities in the early 1930s and made them more susceptible to "wasting disease" (attack by the slime mold *Labyrinthula*) with resultant mass mortality (>90% of standing stocks destroyed).[81] Similar thermal stress effects consequent on global warming could be catastrophic to tropical seagrass communities, although their greater species diversity may make such an eventuality less likely. However, mass mortality of *Thalassia*, beginning in the summer of 1987, has recently been reported in Florida Bay.[82] This had involved complete loss of 4000 ha and damage to over 23,000 ha by 1990. Abnormally high water temperature and attack by the slime mold *Labyrinthula* are among the potential causative factors under investigation. Major losses of shallow seagrass communities would be likely to increase shore erosion, alter sediment transport patterns, and have detrimental effects on adjacent coral reef areas and mangroves.[73,83]

SEA LEVEL RISE, CO_2, AND ULTRAVIOLET RADIATION

Where seawater circulation over dense seagrass beds is restricted, such as in semi-enclosed shallow lagoons, seagrass productivity can be significantly reduced because of nutrient depletion, CO_2 depletion during the day and O_2 depletion at night, and thermal stress due to solar heating. In such situations sea level rise might improve water circulation and thus, productivity. As stated earlier, vertical accretion rates of seagrass beds have been little studied. If they can accrete at the rate quoted for *Posidonia*,[75] then existing dense beds should be able to maintain their depth relative to mean sea level, all other factors being equal. Seagrass communities should also be able to migrate shoreward as sea level rises.

Seagrasses are C_3 plants.[84] Their rates of photosynthesis and growth seem to be limited largely by the rate of supply of carbon as bicarbonate or CO_2.[73,85] This is particularly the case in actively photosynthesizing shallow seagrass beds when tidal currents are slow and pH can rise from 8.2 to 8.9 (or even 9.4) greatly reducing pCO_2 and the availability of bicarbonate. Seagrasses are thus likely to benefit considerably from increased atmospheric and therefore, shallow water CO_2 concentrations, particularly in situations where they may be under other forms of stress as a result of climatic change.

Nothing seems to be known about the likely effects of increased UVB radiation on seagrasses. The major impact of Hurricane Donna on some Florida seagrasses appeared not to be due to violent wave action but to freshwater runoff from land, which led to massive leaf loss but eventual recovery from the more resilient rhizomes.[86] Increased intensity of rainstorms in the humid tropics would be likely to increase incidences of run off of sediment-laden freshwater into shallow lagoons with consequent detrimental effects on seagrass communities therein.

Shallow seagrass communities relying to some extent on coral reefs toward the sea for protection from strong wave action would suffer if the reefs were impacted adversely. Similarly, those relying on landward mangrove systems for protection from the worst effects of terrestrial runoff during heavy rainstorms would suffer if these mangals were badly impacted by climatic change. Such interactive effects between coastal ecosystems may well be far more significant than local effects of climatic variables on seagrasses in isolation.

CLIMATIC CHANGE IN THE CONTEXT OF SHORTER TERM ANTHROPOGENIC DISTURBANCES

Although predicted best-guess rates of sea level rise and global mean temperature rise as a result of greenhouse gas emissions are unprecedented since civilization began, impacts of other human activities on tropical coastal ecosystems[87] may be just as, if not more important over the next half century or so. These more immediate and more localized anthropogenic impacts (Figure 10–6) may also seriously impair the abilities of the ecosystems to respond to the stresses and demands of global climatic change. As a corollary, anything that can be done to relieve these more local impacts will increase the chances of the ecosystems responding optimally to the challenges of climatic change and

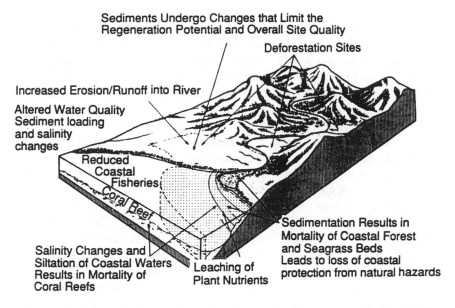

Sediments Undergo Changes that Limit the
Regeneration Potential and Overall Site Quality

Deforestation Sites

Increased Erosion/Runoff into River

Altered Water Quality
Sediment loading
and salinity
changes

Reduced
Coastal
Fisheries

Coral Reef

Salinity Changes and
Siltation of Coastal Waters
Results in Mortality of
Coral Reefs

Leaching of
Plant Nutrients

Sedimentation Results in
Mortality of Coastal Forest
and Seagrass Beds
Leads to loss of coastal
protection from natural hazards

Figure 10–6 Coastal effects of excessive clear-cutting (Snedaker and Getter, 1984).

decrease the long-term negative economic impacts to coastal peoples directly or indirectly dependent on them. This is just one of many reasons for promoting integrated coastal management, which aims at managing systems (rather than activities or single resources) to support multiple uses on a long-term sustainable basis.[88–90] Unwise development focused on single activities has unfortunately often led to a downward spiral of declining resources and increasing conflict with associated poverty and social unrest.

Coral reefs are already under threat from human activities in many areas. Indeed, an international workshop on coral bleaching, coral reef ecosystems, and global change concluded recently that, "Anthropogenic environmental alterations on global, regional and local levels are reason for serious concern about the health and local survival of coral reef ecosystems".[19] Logging and upland agriculture on marginal soils has led to greatly increased erosion of topsoil, which is washed into coastal waters, particularly during heavy rains. The increased sediment load of the coastal waters reduces light reaching coral zooxanthellae (and thus calcification) and causes siltation on reefs, which means coral colonies have to use energy continuously to produce mucus to clear sediment off their surfaces. Combinations of these stresses can lead to widespread coral death in affected areas. Build up of silt may also inhibit settlement of new corals from the plankton. In many countries coral reefs are mined for building material or as a source of lime (e.g., Maldives, Comores, Mozambique, Philippines, and Indonesia), an activity that can involve wholesale destruction of the reef flat. From the standpoint of coastal erosion, removing the top 0.5 m of a protective coral reef by mining it is essentially equivalent to a 0.5-m sea level rise (not forecast for a century!). Chronic oil pollution, such as is seen in the Gulf of Suez, and sewage or thermal discharges can also lead to local coral death. Coral reefs contribute about 12% of the world's fish catch; yet destructive fishing methods, such as blast fishing, muro-ami, cyanide fishing, and kayakas, severely damage reefs, destroying the resource base. It is believed that most coral reef areas within 15 km of the shores of Southeast Asia are now overfished. Such reefs currently provide some 10 to 25% of fish protein available to people living in the coastal zone.[91]

Mangroves may be under threat from accelerated sea level rise over the next century but man is the major threat at present. It was estimated in 1986 that 55% of the original area of mangrove ecosystems in the Afrotropical realm had been lost and 58% in the Indomalayan realm.[92] This has largely resulted from (1) clear-felling of mangrove trees for firewood, charcoal, pulp, rayon, and wood chips, (2) conversion of mangrove areas to aquaculture (e.g., for prawns) and agriculture (e.g., rice and coconuts), and (3) reclamation for housing and industrial developments. In 1970 40% of the total mangrove area of Sabah was allotted for wood-chip export to Japan, while in Indonesia more than 2000 km^2 of mangroves were being exploited for wood chips in the mid-1980s.[83] Mangroves are also very sensitive to oil spills with, for example, hundreds of hectares on the eastern coast of Sumatra suffering dieback after the Showa Maru spill and reports of 98% mortality of mangroves affected by refinery effluent at Cilacap on the southern coast of Java. Another man-made problem for mangrove systems is accelerated subsidence of delta areas due to groundwater extraction (e.g., Bengal Delta[67]) or damming of rivers upstream drastically reducing water, nutrient, and sediment supply (e.g., Indus Delta). Such impacts may be far more important locally than global climatic change; indeed, the once luxuriant Indus mangroves have already been decimated.

Seagrasses have tended to be less impacted by man than mangroves and seem to have suffered most from dredging and filling (reclamation) activities in shallow coastal waters. Short-term effects include direct removal or burial under spoils of plants and long-term effects include adverse impacts of increased sediment in suspension in destabilized areas. The importance of these effects on a global scale has not been quantified. Sewage and industrial wastes including hypersaline outflows from desalination plants and thermal effluents from power plants can also damage seagrass beds.[73]

Thus, the overall picture is of ecosystems, which in many geographical areas are already stressed by man's activities, facing the prospect of further stress from climatic change. Where coral reef or seagrass ecosystems are already close to thermal tolerances during summer, their ability to adapt to climatic change, even if otherwise unimpacted, is by no means clear. If locally impacted by other human activities such ecosystems would seem destined to suffer mass mortalities.

CONCLUSION

Firm data on likely impacts of climate change are largely lacking with important questions from the molecular biochemical level to the eocsystem level unanswered. There are also considerable uncertainties in the predictions of just how climate will change. Of crucial importance to both coral reefs and seagrasses is the question of how fast and by how much sea temperatures in the tropics may rise. As further climatic and biological research addresses the issues now being raised, predictions of

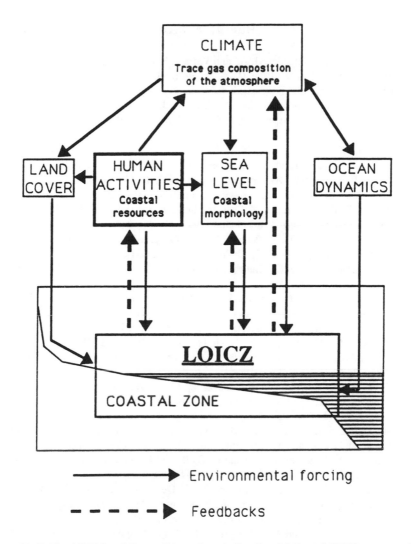

Figure 10-7 The IGBP Land-Ocean Interactions in the Coastal Zone (LOICZ) program. Main effects of global change on the coastal zone, and of changes in the coastal zone on the global environment. (From P. M. Holligan and H. de Boois, 1993, Land-Ocean Interactions in the Coastal Zone (LOICZ). Science Plan, *IGBP Global Change Report* No. 25. With permission.)

impacts will hopefully be refined. One program of research that should make a major contribution to improving our ability to predict impacts of climatic change on coastal ecosystems is the Land-Ocean Interactions in the Coastal Zone (LOICZ) program (Figure 10-7) of the International Geosphere-Biosphere Programme.[93] At present the outlook for the ecosystems and those dependent on them seems somewhat gloomy for the next century, not least because of local anthropogenic impacts.

REFERENCES

1. Houghton, J. T., Callander, B. A., and Varney, S. K., Eds. *Climate Change 1992. The Supplementary Report to the IPCC Scientific Assessment* (Cambridge: Cambridge University Press, 1992), pp. xii, 200.

2. Houghton, J. T., Jenkins, G. J., and Ephraums, J. J., Eds. *Climate Change. The IPCC Scientific Assessment* (Cambridge: Cambridge University Press, 1990), pp. xxxix, 365.

3. Mitchell, J. F. B. and Gregory, J. M. Climatic consequences of emissions and a comparison of IS92a and SA90, in *Climate Change 1992. The Supplementary Report to the IPCC Scientific Assessment,* Houghton, J. T., Callander, B. A., and Varney, S. K., Eds. (Cambridge: Cambridge University Press, 1992), pp. 171–175.

4. Wigley, T. M. L. and Raper, S. C. B. Implications for climate and sea level of revised IPCC scenarios, *Nature* 357:293–300 (1992).

5. Folland, C. K., Karl, T., and Vinnikov, K. Y. Observed climate variations and change, in *Climate Change. The IPCC Scientific Assessment,* Houghton, J. T., Jenkins, G. J., and Ephraums, J. J., Eds. (Cambridge: Cambridge University Press, 1990), pp. 195–238.

6. Warrick, R. A. and Oerlemans, H. Sea level rise, in *Climate Change. The IPCC Scientific Assessment,* Houghton, J. T., Jenkins, G.J., and Ephraums, J. J., Eds. (Cambridge: Cambridge University Press, 1990), pp. 257–281.

7. Melillo, J. M., McGuire, A. D., Kicklighter, D. W., Moore, B., III, Vorosmarty, C. J., and Schloss, A. L. Global climate change and terrestrial net primary production, *Nature* 363:234–240 (1993).

8. Gates, W. L., Rowntree, P. R., and Zeng, Q.-C. Validation of climate models, in *Climate Change. The IPCC Scientific Assessment,* Houghton, J. T., Jenkins, G. J., and Ephraums, J. J., Eds. (Cambridge: Cambridge University Press, 1990), pp. 93–130.

9. Mitchell, J. F. B., Manabe, S., Meleshko, V., and Tokioka, T. Equilibrium climate change and its implications for the future, in *Climate Change. The IPCC Scientific Assessment,* Houghton, J. T., Jenkins, G. J., and Ephraums, J. J., Eds. (Cambridge: Cambridge University Press, 1990), pp. 131–172.

10. Holdgate, M. W., Bruce, J., Camacho, R. F., Desai, N., Mahtab, F. U., Mascarenhas, O., Maunder, W. J., Shihab, H., and Tewungwa, S., *Climate Change: Meeting the Challenge* (London: Commonwealth Secretariat, 1989), pp. ix, 131.

11. Heymsfield, A. J. and Miloshevich, L. M. Limit to greenhouse warming?, *Nature* 351:14–15 (1991).

12. Ramanathan, V. and Collins, W. Thermodynamic regultion of ocean warming by cirrus clouds deduced from observations of the 1987 El Niño, *Nature* 351:27–32 (1991).

13. Gates, W. L., Mitchell, J. F. B., Boer, G. J., Cubasch, U., and Meleshko, V. P. Climate modelling, climate prediction and model validation, in *Climate Change 1992. The Supplementary Report to the IPCC Scientific Assessment,* Houghton, J. T., Callander, B. A., and Varney, S. K., Eds. (Cambridge: Cambridge University Press, 1992), pp. 97–134.

14. Folland, C. K., Karl, T. R., Nicholls, N., Nyenzi, B. S., Parker, D. E., and Vinnikov, K. Y. Observed climate variability and change, in *Climate Change 1992. The Supplementary Report to the IPCC Scientific Assessment,* Houghton, J. T., Callander, B. A., and Varney, S. K., Eds. (Cambridge: Cambridge University Press, 1992), pp. 135–170.

15. Emanuel, K. A. The dependence of hurricane intensity on climate, *Nature* 326:483–485 (1987).

16. Kinsey, D. W. and Hopley, D. The significance of coral reefs as global carbon sinks — response to Greenhouse, *Palaeogeogr. Palaeoclimatol. Palaeontol.* (Global and Planetary Change Section) 89:363–377 (1991).

17. Hopley, D. and Kinsey, D. W. The effects of a rapid short-term sea level rise on the Great Barrier Reef, in *Greenhouse: Planning for Climate Change,* Pearman, G. I., Ed. (Leiden: Brill, 1988), pp. 189–201.

18. Smith, S. V. and Buddemeier, R. W. Global change and coral reef ecosystems, *Annu. Rev. Ecol. Syst.* 23:89–118 (1992).

19. D'Elia, C. F., Buddemeier, R. W., and Smith, S. V. Workshop on Coral Bleaching, Coral Reef Ecosystems and Global Change: Report of Proceedings, Maryland Sea Grant College Publication UM-SG-TS-91-03. (Maryland Sea Grant College, 1991), pp. vi, 49.

20. Buddemeier, R. W. and Smith, S. V. Coral reef growth in an era of rapidly rising sea level: predictions and suggestions for long-term research, *Coral Reefs* 7:51–56 (1988).

21. Neumann, A. C. and MacIntyre, I. Reef response to sea level rise: keep up, catch up or give up?, *Proc. 5th Int. Coral Reef Congr.* 3:105–109 (1985).

22. Fairbanks, R. G. A 17,000 year glacio-eustatic sea level record, *Nature* 342:637–642 (1989).

23. Eakin, C. M. Post-ENSO Panamanian reefs: less accretion, more erosion and damselfish protection, in *7th Int. Coral Reef Symp. Abstr.* Richmond, R. H., Ed. (Guam: University of Guam Marine Laboratory, 1992), p. 27.

24. Smith, S. V. and Kinsey, D. W. Calcium carbonate production, coral reef growth, and sea level change, *Science* 194:937–939 (1976).

25. Coles, S. L., Jokiel, P. L., and Lewis, C. R. Thermal tolerance in tropical versus subtropical Pacific reef corals, *Pac. Sci.* 30:159–166 (1976).

26. Coles, S. L. Limitations on reef coral development in the Arabian Gulf. Temperature or algal competition?, *Proc. 6th Int. Coral Reef Symp.* 3:211–216 (1988).

27. Jokiel, P. L. and Coles, S. J. Response of Hawaiian and other Indo-Pacific reef corals to elevated temperature, *Coral Reefs* 8:155–162 (1990).

28. Glynn, P. W. Widespread mortality and the 1982–1983 El Niño warming event, *Environ. Conserv.* 11:133–146 (1984).

29. Brown, B. E. and Suharsono. Damage and recovery of coral reefs affected by El Niño related seawater warming in the Thousand Islands, Indonesia, *Coral Reefs* 8:163–170 (1990).

30. Goreau, T. J. Coral bleaching in Jamaica, *Nature* 343:417 (1990).

31. Glynn, P. W. and D'Croz, L. Experimental evidence for high temperature stress as the cause of El Niño-coincident coral mortality, *Coral Reefs* 8:181–191 (1990).

32. Goreau, T. J. and Macfarlane, A. H. Reduced growth rate of *Montastrea annularis* following the 1987–1988 coral-bleaching event, *Coral Reefs* 8:211–215 (1990).

33. Szmant, A. M. and Gassman, N. J. The effects of prolonged "bleaching" on the tissue biomass and reproduction of the reef coral *Montastrea annularis, Coral Reefs* 8:217–224 (1990).

34. Glynn, P. W. Coral reef bleaching: ecological perspectives, *Coral Reefs* 12:1–17 (1993).

35. Brown, B. E. and Howard, L. S. Assessing the effects of "stress" on reef corals, *Adv. Mar. Biol.* 22:1–63 (1985).

36. Jokiel, P. L. and Coles, S. L. Effects of heated effluent on hermatypic corals at Kahe Point, Oahu, *Pac. Sci.* 28:1–18 (1974).

37. Cook, C. B., Logan, A., Ward, J., Luckhurst, B., and Berg, C. J., Jr. Elevated temperatures and bleaching on a high-latitude coral reef: the 1988 Bermuda event, *Coral Reefs* 9:45–49 (1990).

38. Parsons, P. A. Conservation and global warming: a problem in biological adaptation to stress, *Ambio* 18:322–325 (1989).

39. Worrest, R. C. and Grant, L. D. Effects of ultraviolet-B radiation in terrestrial plants and marine organisms, in *Ozone Depletion: Health and Environmental Consequences,* Jones, R. R. and Wigley, T., Eds. (Chichester: John Wiley & Sons, 1989), pp. 197–206.

40. Isaksen, I. S. A. The beginnings of a problem, in *Ozone Depletion: Health and Environmental Consequences,* Jones, R. R. and Wigley, T., Eds. (Chichester: John Wiley & Sons, 1989), pp. 15–25.

41. Eckman, R. S. and Pyle, J. A. Numerical modelling of ozone perturbations, in *Ozone Depletion: Health and Environmental Consequences,* Jones, R. R. and Wigley, T., Eds. (Chichester: John Wiley & Sons, 1989), pp. 27–42.

42. Jokiel, P. L. Solar ultraviolet radiation and coral reef epifauna, *Science* 207:1069–1071 (1980).

43. Dunlap, W. C., Chalker, B. E., and Oliver, J. K. Bathymetric adaptations of reef-building corals at Davies Reef, Great Barrier Reef, Australia. III. UV-B absorbing compounds, *J. Exp. Mar. Biol. Ecol.* 104:239–248 (1986).

44. Jokiel, P. L. and York, R. H., Jr. Solar ultraviolet photobiology of the reef coral *Pocillopora damicornis* and symbiotic zooxanthellae, *Bull. Mar. Sci.* 32:301–315 (1982).

45. Siebeck, O. Experimental investigation of UV tolerance in hermatypic corals (Scleractinia), *Mar. Ecol Prog. Ser.* 43:95–103 (1988).

46. Raven, J. A. Responses of aquatic photosynthetic organisms to increased solar UV-B, *J. Photochem. Photobiol. B: Biology* 9:239–244 (1991).

47. Lesser, M. P., Stochaj, W. R., Tapley, D. W., and Shick, J. M. Bleaching in coral reef anthozoans: effects of irradiance, ultraviolet radiation, and temperature on the activities of protective enzymes against active oxygen, *Coral Reefs* 8:225–232 (1990).

48. Smith, S. V. Coral reef area and the contributions of reefs to processes and resources of the world's oceans, *Nature* 273:225–226 (1978).

49. Crossland, C. J., Hatcher, B. G., and Smith, S. V. Role of coral reefs in global ocean production, *Coral Reefs* 10:55–64 (1991).

50. Ware, J. R., Smith, S. V., and Reaka-Kudla, M. L. Coral reefs: sources or sinks of atmospheric CO_2? *Coral Reefs* 11:127–130 (1992).

51. Tans, P. P., Fung, I. Y., and Takahashi, T. Observational constraints on the global atmospheric CO_2 budget, *Science* 247:1431–1438 (1990).

52. Melillo, J. M., Callaghan, T. V., Woodward, F. I., Salati, E., and Sinha, S. K., Effects on ecosystems, in *Climate Change. The IPCC Scientific Assessment,* Houghton, J. T., Jenkins, G. J., and Ephraums, J. J., Eds. (Cambridge: Cambridge University Press, 1990), pp. 283–310.

53. Macnae, W. A general account of the fauna and flora of mangrove swamps in the Indo-West Pacific region, *Adv. Mar. Biol.* 6:73–270 (1968).

54. Ellison, J. C. The Pacific palaeogeography of *Rhizophora mangle* L. (Rhizophoraceae), *Bot. J. Linn. Soc.* 105:271–284 (1991).

55. Hamilton, L. S. and Snedaker, S. C., Eds. *Handbook for Mangrove Area Management* (Honolulu: East-West Center, IUCN and UNESCO, 1984), pp. xii, 123.

56. Woodroffe, C. D. and Grindrod, J. Mangrove biogeography: the role of Quaternary environmental and sea-level change, *J. Biogeogr.* 18:479–492 (1991).

57. Ellison, J. C. and Stoddart, D. R. Mangrove ecosystem collapse during predicted sea-level rise: Holocene analogues and implications, *J. Coastal Res.* 7:151–165 (1991).

58. Ellison, J. C. Pollen analysis of mangrove sediments as a sea-level indicator: assessment from Tongatapu, Tonga, *Palaeogeogr. Palaeoecol. Palaeoclimatol.* 74:327–341 (1989).

59. Scholl, D. W. Recent sedimentary record in mangrove swamps and rise in sea level over the southwestern coast of Florida. I, *Mar. Geol.* 1:344–366 (1964).

60. Woodroffe, C. D. Mangrove swamp stratigraphy and Holocene transgression, Grand Cayman Island, West Indies, *Mar. Geol.* 41:271–294 (1981).

61. Bloom, A. L. Paludal stratigraphy of Truk, Ponape and Kusaie, East Caroline Islands, *Geol. Soc. Am. Bull.* 81:1895–1904 (1970).

62. Ward, J. V. Palynology of Kosrae, Eastern Caroline Islands: recoveries from pollen rain and Holocene deposits, *Rev. Palaeobot. Palynol.* 55:247–271 (1988).

63. Bruun, P. Sea-level rise as a cause of shore erosion, *J. Water. Harbors Div. Proc. Am. Soc. Civ. Eng.* 88:117–130 (1962).

64. Vicente, V. P., Singh, N. C., and Botello, A. V. Ecological implications of potential climate change and sea level rise in the wider Caribbean, in *Implications of Climatic Changes in the Wider Caribbean Region,* Maul, G. A., Ed. UNEP(OCA)/CAR WG.1/INF.3. (1989), pp. 193–223.

65. Pernetta, J. C. and Osborne, P. L. Deltaic floodplains: the mangroves of the Gulf of Papua and the Fly River, Papua New Guinea, in *Implications of Expected Climate Changes in the South Pacific Region: An Overview,* Pernetta, J. C. and Hughes, P. J., Eds. UNEP Regional Seas Reports and Studies 128. (UNEP, 1990), pp. 200–217.

66. Pethick, J. and Spencer, T. Mangrove response to sea level rise: the Rufiji Delta, Tanzania, (n.d.), p. 7.

67. Milliman, J. D., Broadus, J. M., and Gable, F. Environmental and economic implications of rising sea level and subsiding deltas: the Nile and Bengal examples, *Ambio* 18:340–345 (1989).

68. Anonymous. UNEP-IOC-WMO-IUCN Meeting of Experts on a Long-Term Global Monitoring System of Coastal and Near-Shore Phenomena Related to Climate Change, Pilot Projects on Mangroves and Coral Reefs, UNEP-IOC-WMO-IUCN/GCNSMS-II/3. (UNESCO, 1992).

69. Passioura, J. B., Ball, M. C., and Knight, J. H. Mangroves may salinize the soil and in so doing limit their transpiration rate, *Funct. Ecol.* 6:476–481 (1992).

70. Lovelock, C. E., Clough, B. F., and Woodrow, I. E. Distribution and accumulation of ultraviolet-radiation-absorbing compounds in leaves of tropical mangroves, *Planta* 188:143–154 (1992).

71. den Hartog, C. *The Seagrasses of the World* (Amsterdam: North-Holland Publishing Company, 1970), p. 275.

72. Wahbeh, M. I. Seagrasses, in *Red Sea,* Edwards, A. J. and Head, S. M., Eds. Key Environments (Oxford: Pergamon Press, 1987), pp. 184–191.

73. Zieman, J. C. The ecology of the seagrasses of South Florida: a community profile, FWB/OBS-82/25. (US Department of the Interior, Bureau of Land Management Fish and Wildlife Service, 1982), pp. viii, 123.

74. Burrell, D. C. and Schubel, J. R. Seagrass ecosystem oceanography, in *Seagrass Ecosystems: A Scientific Perspective,* McRoy, C. P. and Helfferich, C., Eds. (New York: M. Dekker, 1977), pp. 195–232.

75. Elder, D. and Jeudy de Grissac, A. Effects of the sea level rise on coastal ecosystems including those under special protection and threatened and migratory species, UNEP(OCA)/WG.2/5. (UNEP, 1988), p. 35.

76. Ward, L. G., Kemp, W. M., and Boynton, W. R. The influence of waves and seagrass communities on suspended particulates in an estuarine environment, *Mar. Geol.* 59:85–103 (1984).

77. Fortes, M. D. *Seagrasses: A Resource Unknown in the ASEAN Region* (Manila: ICLARM, 1989), pp. x, 46.

78. Bell, J. D. and Pollard, D. A. Ecology of fish assemblages and fisheries associated with seagrasses, in *Biology of Seagrasses,* Larkum, A. W. D., McComb, A. J., and Shepherd, S. A., Eds. (Amsterdam: Elsevier, 1989), pp. 565–609.

79. Lanyon, J., Limpus, C. J., and Marsh, H. Dugongs and turtles: grazers in the seagrass ecosystem, in *Biology of Seagrasses,* Larkum, A. W. D., McComb, A. J., and Shepherd, S. A., Eds. (Amsterdam: Elsevier, 1989), pp. 610–634.

80. McRoy, C. P. and McMillan, G. Production ecology and physiology of seagrasses, in *Seagrass Ecosystems: A Scientific Perspective,* McRoy, C. P. and Helfferich, C., Eds. (New York: M. Dekker, 1977), pp. 53–88.

81. Rasmussen, E. The wasting disease of eelgrass (*Zostera marina*) and its effects on environmental factors fauna, in *Seagrass Ecosystems: A Scientific Perspective,* McRoy, C. P. and Elfferich, C., Eds. (New York: M. Dekker, 1977), pp. 1–51.

82. Robblee, M. B. et al. Mass mortality of the tropical seagrass *Thalassia testudinum* in Florida Bay (USA), *Mar. Ecol. Prog. Ser.* 71:297–299 (1991).

83. Fortes, M. D. Mangrove and seagrass beds of East Asia: habitats under stress, *Ambio* 17:207–213 (1988).

84. Andrews, T. J. and Abel, K. M. Photosynthetic carbon metabolism in seagrasses (^{14}C-labelling evidence for the C_3 pathway), *Plant Physiol.* 63:650–656 (1979).

85. Beer, S. and Waisel, Y. Some photosynthetic carbon fixation properties of seagrasses, *Aquat. Bot.* 7:129–138 (1979).

86. Thomas, L. P., Moore, D. R., and Work, R. C. Effects of Hurricane Donna on the turtle grass beds of Biscayne Bay, Florida, *Bull. Mar. Sci. Gulf Caribb.* 11:191–197 (1961).

87. Gomez, E. D. Overview of environmental problems in the East Asian seas region, *Ambio* 17:166–169 (1988).

88. Clark, J. R. *Integrated Management of Coastal Zones* (Rome: FAO, 1992), pp. viii, 167.

89. Burbridge, P. R. Coastal and marine resource management in the Strait of Malacca, *Ambio* 17:170–177 (1988).

90. Edwards, A. J. Managing systems not uses: the challenges of water-borne interdependence and coastal dynamics, in *Priorities for Water Resources Allocation and Management,* (London: Overseas Development Administration, 1993), pp. 149–154.

91. McManus, J. W. Coral reefs of the ASEAN region: status and management, *Ambio* 17:189–193 (1988).

92. World Resources Institute. *World Resources 1990–1991* (New York and Oxford: Oxford University Press, 1990), pp. xiv, 383.

93. Holligan, P. M. and de Boois, H. Land-Ocean Interactions in the Coastal Zone (LOICZ). Science Plan, IGBP Global Change Report No. 25. (IGBP, 1993), p. 50.

94. Bijlsma, L., O'Callaghan, J., Hillen, R., Misdorp, R., Mieremet, B., Ries, K., Spradley, J. R., and Titus, J., Eds. *Global Climate Change and the Rising Challenge of the Sea. Report of the Coastal Zone Management Subgroup* (The Hague: Ministry of Transport, Public Works and Water Management, 1992).

CHAPTER 11

Impact of Climatic Change on Coastal Agriculture*

R. Brinkman

INTRODUCTION

The expected gradual but increasingly rapid sea level rise, estimated to reach between 0.3 and 1 m by the year 2100 is the main factor specifically affecting agriculture in low-lying coastal areas. The various effects consequent on a gradual sea level rise are discussed below. It should be noted, however, that integrated management of coastal zones should also take into account the various worldwide effects of climate change, indicated in the next section.

A sea level rise of, for example, 0.6 m, which could affect some land even up to several meters elevation by storm-surge flooding and saltwater intrusion up river estuaries, could potentially affect several million square kilometers, or up to about 3% of the world's total land area.[1] The coastal lowlands potentially at risk include about one third of the world's total cropland and have a present population of about 1 billion — a figure that might be doubled within the next 100 years. Thus, the potential effects of a rise in sea level could be very much greater than would be suggested by the small proportion of the world's land and soils that might be affected.

It is important to recognize that these potential figures grossly exaggerate the soil areas and population that might actually be at risk. The scenario of sea level rising contour by contour to overwhelm, for example, 20% of Bangladesh and the Nile Delta,[2] ignores dynamic changes that will be taking place in coastal lowlands concurrently with, or in response to a gradual rise in sea level. Some of these changes are natural, such as continuing or accelerated sediment deposition in some deltas and estuary areas, which would reduce or totally offset the effects of a rise in sea level.

---

* This chapter was prepared by Robert Brinkman, Land and Water Development Division of the Food and Agriculture Organization of the United Nations (FAO), and is reproduced here by permission of that organization.

Other changes are the result of human action, such as the building of embankments to protect land and people from the effects of a rising sea level.

In most of the world, coastal lowlands used for agriculture are still under the influence of the natural sedimentary and hydrologic conditions. These show major variations from place to place, both within any one coastal plain or delta as well as between different ones. With a gradual sea level rise, the response of coastal lowlands will be similarly variable. A higher sea level does not merely translate into coastal erosion, salt intrusion, and higher water levels some distance inland. With the sea level rising, a larger proportion of sediment brought down by rivers may be deposited on river levees and in the margins of backswamps, for example, while less would reach the river mouths, to be redistributed along the coast or on an undersea delta. The land on the levees could thus remain very shallowly flooded, with similar potential for agriculture as before, while the backswamps might become gradually more deeply flooded.

Little attention appears to have been given to date to soil changes that might take place with a climatically induced rise in sea level. However, prediction of the likely consequences on soil properties and productivity need not be entirely speculative. There is abundant geological, historical, and contemporary evidence of the effects of a rise in sea level on adjoining land areas. Examples include:

- The effects of the 100-m eustatic rise in sea level following the last glacial period on the coastal areas bordering the North Sea as described by Jelgersma[3]
- Land subsidence in relation to sea level of about 1 m during the last 2000 years in the Netherlands, attributed to tectonic subsidence of the North Sea basin and eustatic sea level rise by geoidal changes following the last glacial period[3]
- Geological and archaeological evidence of land subsidence in the Mississippi, Nile, Rhine, and other deltas resulting from tectonic warping or compaction of alluvial sediments[3-5]
- Recent evidence of local land subsidence occurring as a consequence of water, natural gas, or oil extraction in the northern Netherlands, Venice, Bangkok, Osaka, the Mississippi Delta, and elsewhere[6-8]
- Worldwide evidence of a 10- to 15-cm eustatic rise in sea level during the last 100 years[1]

There is also evidence from a number of places that human interventions could aggravate the effects of a rising sea level: e.g., by interrupting sediment supply to deltas by dams across rivers upstream, as has happened in the Nile Delta following the construction of the Aswan High Dam in 1964;[9] and by embanking rivers to channel flood flows directly to the sea, as in the Mississippi Delta, thus depriving basin areas of sediment supplies which formerly offset land subsidence due to compaction of the deltaic sediments.[10] In several localities, rates of land subsidence by natural causes or directly attributable to human activities are much greater than the rates of sea level rise estimated to occur in response to global warming during the next century.

WORLDWIDE EFFECTS

Several aspects of human-induced climate change are expected to affect agriculture worldwide. Higher CO_2 concentrations would stimulate photosynthesis and increase water-use efficiency of crops and natural vegetation, with positive effects on biomass accumulation and crop yields. Higher temperatures, especially at night, would increase respiration rates, which would counteract these positive effects. The higher temperatures and evaporation rates may increase cloudiness, also with a negative effect on yields because of decreased photosynthetically active radiation and because of possible increased incidence of fungal diseases.

Greater rainfall variability may increase drought hazard. Frost-free periods may become longer and lengths of growing periods as determined by moisture availability may become longer or shorter than at present in different parts of the world.

Natural selection pressure may be expected to favor varieties — and plant breeding to provide cultivars, selected varieties — that are more adapted to the changed conditions, so that adverse effects are minimized and the benefits of potentially favorable changes are captured.

Besides direct effects on crops and natural vegetation, climate change may influence to some extent the soils on which they grow. Soil temperature increases would influence organic matter dynamics. They would reduce the extent of permafrost while increasing soil reduction in high latitudes. Increased variability and more high-intensity rainfall events would entail higher leaching rates in most soils, with more runoff on slopes and consequent erosion and sedimentation; in humid climates, increased areas with periodic water saturation and soil reduction; in arid areas, salinization.

Soils with the most resilience against degradation by these and other processes have a high infiltration rate and structural stability, and moderate or good external drainage. Management measures to increase soil resilience against the effects of climate change should increase vegetative cover and activity of soil fauna, the same measures as against other perturbations and stresses caused by human exploits. Direct human impacts, for example through overgrazing, plant nutrient mining by arable farming with very low external inputs, or pollution by excessive applications of manure or pesticides are generally more damaging, and more immediately so, than the effects of gradual climate change on soils.

AGRICULTURE IN COASTAL LOWLANDS

Some coastal lowlands have been strongly modified by human action, with heavy investment in dikes and drainage systems. Only in such areas is agriculture largely free of the influence of a sea level rise, except for a minor increase in costs for drainage pumping and a moderate increase in frequency of storm-flood hazards. This may require raising dikes and making them heavier in due course, perhaps in the first decades of the next century.

Many coastal lowlands, especially in the developing world, have no or limited human-built protection against nature's variations in space and time. There, agriculture and other land uses are closely tuned to the existing differences in elevation of the land surface, in drainage and normal seasonal flooding conditions, and in flood and cyclone hazards.

In a tropical coastal plain one may encounter, for example, a broad belt of mangrove forest along the coast and other saline and brackish tidewater, with different mangrove species growing, or planted, depending on local salinity and tidal range and used for different purposes. Parts of this forest may have been cleared and the land converted into complexes of fish or shrimp ponds, or into salt pans where the dry season is long. These land uses may border on the coast in protected places or remain behind a protective mangrove strip where cyclones or other storm floods are prevalent.

Inland from the mangrove belt, the land is nonsaline, or seasonally desalinized by the monsoon rains. There, a single crop of wetland rice could be found, grown in bunded fields, ploughed and puddled after the early monsoon rains. Normal flooding depth is small, and short-statured, highly productive cultivars can be grown if the soil fertility is high or if moderate amounts of fertilizers can be applied. More than one rice crop could be grown where the climate is equatorial (with a very long moisture-determined growing period) or where irrigation water is available. This would normally have to be conveyed from an offtake further inland, since local surface water could be tidal and brackish at least in the dry season, and local groundwater brackish or in short supply. Narrow slightly higher strips of land (actual or former tidal levees) and patches of artificially raised land would be used as homesteads, village sites, and home gardens or fields for the dryland crops needed by the population.

In an equatorial climate, with or without a short dry season, there may be large level peat areas inland from the shallowly flooded riceland. The land surface is slightly above sea level, and the peat overlies clay at depths between a few decimeters and a few meters. Near the centers of interfluves in this flat coastal landscape, far from the rivers, there are peat domes, grown up to some 4 to 10 m above sea level, with flat tops and very gently sloping margins to the surrounding level peat, and with a very shallow radial drainage pattern.

Depending on local population pressure and investment possibilities, the peat areas have remained under swamp forest, or have been converted into large oil-palm plantations or pineapple fields, both with gravity drainage to tidewater, as in parts of Malaysia. Areas of very shallow peat over clay are being used by small farmers for wetland rice cultivation and a range of dryland crops, including fruits and vegetables in and near homesteads, or coconuts on high mounds or ridges built in the rice fields, as in parts of Kalimantan, Indonesia. Large peat areas may also be enclosed by a dike and used to conserve water for irrigation during the short dry season on land nearer the coast, as in Guyana.

In monsoon climates, with their alternating wet and dry seasons, peat areas are smaller or absent. There, the low-lying land inland from tidal influence consists of river levees, actual and abandoned, and generally large backswamp areas with clayey soils, seasonally flooded to moderate or, further inland, considerable depth.

In Bangladesh, for example, the highest parts of river levees are just above water or very shallowly flooded at the height of the rainy season or of the seasonal river flood. At that time, backswamps in the central part of the country are flooded up to 1 to 3 m, and in the northeastern part up to some 5 m depth. During the wet season, shallowly flooded land is used for a main crop of short-statured, high-yielding wetland rice or prime quality wetland rice or jute. Lower-yielding, tall, traditional rice varieties are grown where floods are up to about 1 m; still lower-yielding "floating" rice varieties are grown where floods are up to some 2 to 4 m depth with floodwaters rising slowly, less than 5 cm/day. Deepwater areas with rapidly or irregularly rising floodwater cannot be used for crops in the wet season.

Shallowly flooded land may be used for an additional rice crop during the early rainy season, before the floods and the main wetland crop. After the main crop, much land is used for oilseeds or pulses, sown just after the floods have receded, and increasing areas are producing high-yielding rice in the dry season, irrigated from local surface water or from tubewells.

In tropical arid or semiarid climates, low-lying coastal areas generally have narrow bands of mangrove vegetation along tidewater; in very arid environments, there may be as little as a single line of low trees. On the landward side, this is bordered by dry, hypersaline, barren or virtually barren land, which is shallowly flooded by seawater during occasional spring tides or rare storm surges. Much of this water subsequently evaporates, adding further amounts of salts to the surface and the upper soil layers. Except for occasional salt pans, this land cannot be used for production; some wood is harvested from the mangrove strip.

EFFECTS OF SEA LEVEL RISE ON AGRICULTURE IN COASTAL LOWLANDS

A gradual sea level rise may cause a complex of changes in different parts of an estuary or coastal plain, depending on a range of external factors and on the nature of the area itself. These factors include the length of the dry season, the tidal range, the incidence of cyclonic storms and storm surges, the exposure of the coastline and the nature of the shelf offshore, and the supply of sandy or fine-textured sediments to the area by longshore currents or from rivers.

The nature and extent of the changes also depend on the nature of the sediments (clayey, sandy, peat) in the coastal area itself, on the closed or open nature of the coast and the presence of coastal ridges or dunes, on the presence and density of any network of tidal distributaries, and on the presence of any coastal defenses such as seawalls, or other modifications to the natural hydrology by human engineering, such as polder embankments.

With a rise in sea level, saltwater intrusion may reach intake structures for irrigation water on rivers in coastal plains or deltas, especially in the dry season when a freshwater supply is most needed. Drainage of agricultural land by gravity to rivers may become slower or impossible even upstream from any saltwater influence.

It should be noted that such effects are not specific to accelerated sea level rise. Saltwater intrusion may also occur because dry-season river flows are diminished by abstraction of irrigation water upstream, or because large-scale pumping of ground-water has led to appreciable subsidence, as in parts of the Bangkok Plain. Such human-induced subsidence may also affect drainage of agricultural land, as in the northern coastal plain in the Netherlands. There, the progressive depletion of a large natural gas field has caused up to about 0.3 m subsidence, requiring modification of the drainage system to maintain its efficacy. Human activities that may aggravate, or counteract, the effects of sea level rise are discussed in the last section of this chapter.

Increased frequency of storm surges combined with rising mean sea level may cause lateral erosion of exposed coasts, with landward movement of any sand, shell, or clayey beach ridges. Direct effects of a gradual sea level rise on agriculture in the coastal strip are described in the next section. Changes inland in saline, brackish-water, and freshwater tidal zones and the lower parts of river plains will vary with the climatic conditions and are described separately for monsoon and equatorial climates in the following sections.

DIRECT EFFECTS OF SEA LEVEL RISE NEAR THE COAST

The most direct effects of a rise in sea level on agriculture can be expected in brackish-water aquaculture. Many coastal lowlands in the tropics are fringed by a band of brackish-water fish and shrimp ponds, excavated in former mangrove areas. These are linked to the sea or to estuary channels by a tidal canal or ditch. The ponds are filled, or water is renewed, by opening a sluicegate in the dike surrounding the pond at high tide and they are emptied during harvest by opening the gate at low tide, after insertion of a net or a grill in the opening to retain the fish. Any increase in the level of low tide would slow down or prevent emptying of the ponds, and a moderate increase in spring tide levels (highest) would overtop the sluicegates, leading to loss of fish or shrimp to open tidewater. Damage or loss due to storm floods would also tend to become more frequent with a rise in sea level. Very productive fish and shrimp pond areas could in due course be protected by raising the outer dikes and by installing a complete system of water intake and pumped drainage.

Salt pans are also very sensitive to sea level changes, but since they do not need to drain to tidewater, in contrast to fish or shrimp ponds, protection against gradually higher sea levels would be less costly.

Coastal agriculture also includes the management of mangrove forests on land subject to tidal flooding along the coast and estuaries. Products are timber and fuelwood, charcoal, and a range of minor forest products, for example tannin, honey, and molluscs collected from aerial mangrove roots exposed at low tide.

Such tidal forest areas can survive slow subsidence of a few millimeters per year, as is common in coastal lowlands, if the sediment supply to the land is sufficient to keep the land surface near mean sea level. In many places however, the sediment supply may not be enough to keep up with an effective doubling or tripling of the rate of relative sea level rise. The resulting coastal erosion and water stagnation and deeper, virtually permanent flooding of the more inland parts of the forest would damage or destroy parts of the habitat and the agricultural activities depending on them.

In arid climates, as in Senegal, intrusion of seawater in the hypersaline, barren land behind the mangrove strip may partly desalinize a part of the salt flat, so that the mangrove strip would gradually become somewhat wider. Also, the productivity of the existing mangrove strip could improve to some extent because of the decrease in salinity through increased tidal flushing.

A dynamic equilibrium between sedimentation and a gradual relative sea level rise (including subsidence) can also be broken by other, more immediate, causes than a gradual sea level rise accelerated by human-induced global warming. The many major dams and reservoirs constructed on the world's rivers have reduced, and in some cases virtually stopped, the sediment supply to the deltas and, hence, to the adjacent coastline. This has already been causing coastal erosion and damage to mangrove forests in some coastal lowlands.

CHANGES IN SEASONALLY WET-DRY CLIMATES

In the tidal or estuarine floodplains, storm surges may bring intermittent saltwater flooding, inducing soil salinization. Spring-tide flooding during a rising sea level may also cause seasonal salinization, particularly in the dry season or near equinoxes.

In a broad band along saline or brackish tidewater, the natural rates of vertical accretion will be increased by a slow rise in sea level, the rates also depending on the sediment supply, storm surge incidence, and extent of spring-tide flooding. In these areas, as well as in the clay plains further away from tidewater, the increased incidence and duration of brackish-water flooding will also extend and intensify reducing conditions in soils. This increasing incidence of salinity would decrease the proportion of land suitable for rain-fed wetland rice in such areas, and reduce the available length of the growing period in the remainder, which would tend to decrease potential yields.

In the widening belt influenced by brackish-water flooding, fresh sediment, and increased reduction, mangrove may begin to dominate again (generally *Rhizophora* species in the tropics; or reeds such as *Phragmites* in temperate climates). These vegetation types cause increased accumulation rates of soil organic matter and, in combination with sulfate from the brackish water, accumulation of pyrite and organic sulfur compounds. Potential acid sulfate conditions thus may become more widespread.

With a sea level rise, the slowing down of river flows would entail higher sedimentation rates in riverbeds and consequent higher river flood levels even inland from the tide-influenced lower courses. The frequency of flood damage to agriculture on the adjacent levees would increase, but increased sedimentation on levees along main rivers would be expected to keep pace with the gradual rise in river flood levels.

The increasing flooding depths in the backswamps would not be fully compensated by increasing sediment supply, and the extent of land suitable for the higher-value or higher-yield crops and varieties would shrink, with an increase in the extent of land only suitable for floating rice. There would also be less time available for dryland crops between flood seasons. Soils subject to freshwater flooding or seasonal saturation will be reduced for longer periods and their organic matter contents will tend to increase. Land that becomes perennially wet will be liable to accumulation of a peat cover, or existing peat areas will increase in thickness and extent. Such changes, unless counteracted by protective embankments or artificial drainage, would generally reduce soil productivity or potential. Specifically:

- Land suitability of levee soils for perennial dryland crops would decrease
- Areas suitable for annual dryland crops would decrease and this land might become better suited for wetland crops such as rice or jute
- Areas presently shallowly flooded and well suited to high-yielding wetland rice cultivars might only be suited for traditional, tall varieties; and
- Land now deeply flooded but dry for part of the dry season, and used for floating rice and perhaps a short-season oilseed crop, may become permanently flooded and unsuitable for cultivation

CHANGES IN HUMID CLIMATES WITHOUT OR WITH A SHORT DRY SEASON

In humid climates without a significant dry season, a gradual sea level rise would tend to widen the strip influenced by brackish-water flooding and may give rise to gradual vertical accretion of fine-textured sediment. Such areas would be colonized by mainly *Rhizophora* mangrove and large contents of pyrite would gradually accumulate in the soil. Much of the newly deposited material would thus become potential acid sulfate soil. This would be at the expense of land presently suitable, or actually used, for wetland rice, coconuts, oil palm, sugarcane, pineapple, or rubber.

Further from tidewater, peat growth would tend to accelerate under natural conditions and the peat surface would thus keep pace with the rising groundwater level, linked to the rising mean sea level. In peat areas presently used for estate crops such as pineapple or oil palm, gravity drainage to tidewater would initially become less effective and more difficult to manage, and with further increase in sea level, pump drainage may become necessary to maintain adequate drainage depth. The increasing costs might eventually lead to abandonment, especially of land under tree crops. The land would then revert to swamp vegetation under which peat growth would resume.

Where the protective strip of mangrove forest is removed or is inadequate in width, higher storm surges associated with a higher sea level could inject enormous volumes of seawater onto the peat land. This would kill the freshwater vegetation in the area affected. The peat would then become highly vulnerable to wave erosion and open water might form. This would generally be a shallow freshwater lake, subsequently invaded by a floating herbaceous vegetation and, in the long run, recolonized by freshwater swamp forest. A breach in the strip of land separating the lake from the coast or from a wide tidal channel would, however, convert it into a saline or brackish lagoon and thus change the coastline.

HUMAN ACTIVITIES AGGRAVATING OR COUNTERACTING THE EFFECTS OF SEA LEVEL RISE

Human activities may influence the response of low-lying coastal areas to sea level rise in different ways. The interventions may be remote, as in the case of dam construction, or local, as in the case of subsidence because of drainage.

The construction of dams in upstream areas, for hydroelectric power, irrigation supplies, or protection of land downstream from river floods, blocks sediment supply downstream and to coastal areas. This prevents or slows down vertical accretion, thus further aggravating saltwater intrusion and impairing drainage conditions in riverine, deltaic, or estuarine areas. It also diminishes or blocks sediment supply to the coast itself, which may give rise to retreat of the coastline by wave erosion and longshore transport of the eroded material, as has occurred in the Nile Delta following the completion (1964) of the Aswan High Dam.[9] In the northeastern Nile Delta, the problem of sea level rise would be compounded by the ongoing, probably neotectonic, subsidence of about 0.5 m/century estimated by Stanley;[5] until 1964, this subsidence was compensated by the deposition of sediments from the annual Nile floods.

Ongoing natural, geological erosion, or human-induced erosion caused by large-scale deforestation, constitute major sources of sediment for low-lying coastal areas. In the Netherlands, the upper 4 m of mainly clayey sediment in coastal areas, deposited over the last 4000 years, largely originated from central European sources. This sediment supply prevented major changes in the coastline during the slow sea level rise of about 4 m over the same period. Major soil conservation activities over large upland areas could therefore have effects on coastal lowlands similar to those of large dams.

River draining works, such as are under consideration for the Brahmaputra River in Bangladesh, can increase the flow of water and sediment through the main channel at high river stages. This tends to increase the sediment supply and deposition downstream, which could partly offset the effects of a rising sea level.

Increased water abstraction for irrigation development in upstream areas, particularly from rivers during the dry season, tends to increase the distance of seawater penetration in watercourses in coastal areas. This would aggravate the effects of sea level rise in cutting off existing irrigation inlets from freshwater supplies and by extending the area of coastal saline soils.

Pumping of groundwater or natural gas or oil in a low-lying coastal area may accelerate land subsidence to rates up to several times those envisaged for the sea level rise over the next century. Natural gas extraction, for example, has lowered part of the northern provinces in the Netherlands by about 0.3 m over about 30 years; parts of the Mississippi Delta are subsiding more rapidly than in the 19th century, probably in part because of oil and gas extraction; and parts of Bangkok City have subsided by more than 1 m, and are currently subsiding by about 0.12 m/year, by large-scale pumping of water for urban and industrial supply. Parts of Taipei have subsided more than 2 m during the last 20 years for the same reasons.[11] This subsidence is creating serious land drainage and saline intrusion problems in affected areas.

Groundwater extraction from coastal aquifers for irrigation or other uses may also induce seawater intrusion into these aquifers at rates far exceeding those envisaged as a result of sea level rise alone, particularly in arid and semiarid climates. Sea level rise would therefore compound an already growing problem.

The fresh groundwater in coastal dunes and coral islands often occurs in the form of a lens, in dynamic equilibrium with approximate vertical isostasy at the boundary between fresh and saline water. This requires that the groundwater level be above mean sea level by an amount of about one fortieth of the total thickness of the freshwater lens. If the sea level rises, the volume of the freshwater lens would remain dynamically stable only if its surface would also rise by a similar amount. If the groundwater surface were to remain constant, the depth of freshwater storage would be reduced in due course by up to 40 times the sea level rise, which could drastically reduce or eliminate local water supplies in coral islands and coastal dune areas. A narrowing of dunes or coral islands by coastal erosion would entail similarly serious consequences for the capacity of a freshwater lens.

Low-lying coastal areas normally subside by slow compaction, which can be compensated by vertical accretion of fresh sediment. The Mississippi Delta, for example, was subject to compaction and sedimentation at a rate of about 1 m per century before human influence became dominant.[12] Embankment against seawater intrusion or freshwater flooding stops vertical accretion in the embanked area, but subsidence by compaction or by oxidation of peat continues, so that drainage conditions tend to deteriorate. Eventually, this leads to loss of land to shallow freshwater lakes in basin areas.

Artificial drainage of coastal soils by ditches or subsurface drains, with evacuation of the drainage water through sluice gates at low tide or by low-lift pumping, will accelerate land subsidence. This might be very rapid in peat soils, which are compacted as well as oxidized at high rates after drainage and air entry, but it can occur also in clayey soils. This entails a need for progressively higher embankments and greater pumping heads of the drainage water, with consequently increasing capital and running costs, aggravating the effects of a sea level rise.

REFERENCES

1. Hekstra, G. P. 1989. Global warming and rising sea levels: the policy implications. *Ecologist* 19(1): 4–15.
2. Broadus, J., J. Milliman, S. Edwards, D. Aubrey, and F. Gable. 1986. Rising sea level and damming of rivers: possible effects in Egypt and Bangladesh. pp. 165–189 in: Titus, J. G., Ed. *Effects of Changes in Stratospheric Ozone and Global Climate*. Vol. 4: Sea level rise. UNEP/U.S.EPA.

3. Jelgersma, S. 1988. A future sea level rise: its impacts on coastal lowlands. pp. 61–81 in: *Geology and Urban Development. Atlas of Urban Geology*, Vol. 1. UN-ESCAP.

4. Morgan, J. P. 1970. Depositional processes and products in the deltaic environment. pp. 31–47 in: Morgan, J. P. and R. H. Shaver, Eds. *Deltaic Sedimentation, Modern and Ancient*. Special Publ. No. 15, Society of Economic Paleontologists and Mineralogists, Tulsa, Oklahoma.

5. Stanley, D. J. 1988. Subsidence in the Northeastern Nile delta: rapid rates, possible causes, and consequences. *Science* 240:497–500.

6. Jelgersma, S. 1992 (in press). Land subsidence in coastal lowlands. in: *Proc. SCOPE Workshop Rising Sea level Subsiding Coastal Areas*. Bangkok, Nov. 9–13 1988. SCOPE Vol. nn, Wiley.

7. UNEP 1988. Report of the joint meeting of the task team on implications of climatic changes in the Mediterranean and the co-ordinators of task teams for the Caribbean, Southeast Pacific, South Pacific, East Asian seas and South Asian seas regions, Split, 3–8 October 1988. UNEP(OCA)/WG.2/25.

8. Volker, A. 1989. Impacts of climatic changes on hydrology and water resources of coastal zones. Vol. 1, p. 114–127 in: *Conf. Climate Water*, Helsinki 11–15 Sept. 1989. Publ. Acad. Finland, Govt. Printing Centre, Helsinki.

9. Coutellier, V. and D. J. Stanley. 1987. Late Quaternary stratigraphy and paleogeography of the eastern Nile delta, Egypt. *Mar. Geol.* 77:257–275.

10. Day, J. W. and P. H. Templet. 1989. Consequences of sea level rise: implications from the Mississippi delta. *Coastal Management* 17:241–257.

11. Poland, J. F. Ed., 1984. Guidebook to studies of land subsidence due to groundwater withdrawal. *Studies and Reports in Hydrology* No. 40. Unesco, Paris.

12. Titus, J. G. 1987. Causes and effects of sea level rise. pp. 125–139 in: *Preparing for Climate Change*. Proc. 1st N. Am. Conf. Preparing Climate Change: Cooperative Approach. October 27–29 1987, Climate Institute, Washington, D.C. Government Institutes, Inc., Rockville, MD.

Index